This is Hemp Paper / What better way to demonstrate the versatility of hemp than by printing this book on hemp
paper, using hemp-oil-based inks, all derived from locally grown
hemp? While theoretically possible, the infrastructure for supplying the demand for hemp products is not yet developed to the point
where hemp represents an economical choice for a product such as
this book. In choosing materials for the manufacture of *Hemp
Horizons* our decision-making has balanced idealism and practicality to come up with a solution that tells a positive story of present-
day environmental options.

These endpapers are printed on 100-percent, 54-pound hemp
text paper. You can see that it is a beautiful sheet. It is simply not
yet produced in quantities that make it a viable alternative for the
entire book. Choosing 100-percent hemp papers would have more
than doubled the cover price of the finished book, which would
have conflicted with our desire to have the hemp story read as
widely as possible.

The book's text is printed on 70-pound Roland New Life
Opaque, a recycled sheet containing 30-percent post-consumer
material. Sheets with higher content sacrifice print and archival
quality. Several participants in the production of *Hemp Horizons*
lobbied to print the book on 100-percent tree-free paper. We did
significant research on papers made with kenaf, bamboo, and various combinations using agricultural waste, such as kenaf/wheat
straw. Many look promising, but questions of cost, availability,
lead-time, and reliability made them risky choices for a job of this
magnitude.

The cover is printed on recycled stock, but is protected by a lam-
inate for which there is no commercially available, benign substitute. Booksellers and book buyers demand books that resist
scratching and that maintain a bright appearance for many years.
They should not have to lower their expectations or standards.
Instead, it is incumbent on those of us who trade in the written
word to insist on products that better reflect our goals of sustainability. Perhaps one of today's new hemp entrepreneurs looking for
an opportunity will recognize the area of eco-laminates as a market
niche in need of a good product.

Hemp Horizons

Real Goods Trading Company in Ukiah, California, was founded in 1978 to make available new tools to help people live self-sufficiently and sustainably. Through seasonal catalogs, a periodical (*Real Goods Renewables*), a bi-annual *Solar Living Sourcebook*, as well as retail outlets, Real Goods provides a broad range of tools for independent living.

"Knowledge is our most important product" is the Real Goods motto. To further its mission, Real Goods has joined with Chelsea Green Publishing Company to co-create and co-publish the Real Goods Solar Living book series. The titles in this series are written by pioneering individuals who have firsthand experience in using innovative technology to live lightly on the planet. Chelsea Green books are both practical and inspirational, and they enlarge our view of what is possible as we enter the next millennium.

Ian Baldwin, Jr.	*John Schaeffer*
President, Chelsea Green	President, Real Goods

HEMP
HORIZONS

The Comeback of the World's

Most Promising Plant

JOHN W. ROULAC
and HEMPTECH

CONTRA COSTA COUNTY LIBRARY

Chelsea Green Publishing Company

WHITE RIVER JUNCTION, VERMONT

Book design by Christopher Kuntze.

Printed in Canada
99 98 97   1   2   3   4   5

LIBRARY OF CONGRESS CATALOGING-IN-PUBLICATION DATA

Roulac, John, 1959–
      Hemp horizons : the comeback of the world's most promising plant /
John Roulac and Hemptech.
                        p.   cm.
      Includes bibliographical references (p. 191) and index.
      ISBN 0–930031–93–8
      1. Hemp   2. Hemp industry   I. Hemptech (Firm)   II. Title.
SB255.R68   1997
338.1'7353—dc21   97–18376

Chelsea Green Publishing Company
P.O. Box 428
White River Junction, VT 05001
Telephone: (800) 639-4099 or (603) 448-0317
www.chelseagreen.com

CONTENTS

FOREWORD

As we move into the twenty-first century, American agriculture faces enormous challenges and opportunities. New crops will be vital to the profitability and sustainability of America's land resources, and the cultivation and utilization of industrial hemp could yield substantial benefits to American farmers, manufacturers, and consumers.

The early years of the next century must be used to revitalize America's basic strengths in pursuing solutions to major problems, Among our priority concerns are the trade deficit, job creation, environmental issues, and the globalization of the economy, which is stressing America's working class. And sustainability is becoming a central theme in our vision for the future: America must move into new industries and aggressively develop existing industries that credit nature and at the same time benefit society. Industrial hemp can play an important role in meeting these challenges.

Industrial hemp can generally be grown without pesticides, and can be grown in rotation with nearly all agricultural crops. Hemp can also displace raw materials such as wood-, cotton-, and petrochemical-based fibers, which have higher environmental costs of production, use, and disposal. *Hemp Horizons* details the immense value of hemp, and clarifies its non-drug nature, while celebrating hemp's positive impact on rural development, the environment, and sustainable industry.

The world is facing a major fiber crisis in the coming decade. For example, Wisconsin paper mills must now import from out of our state approximately 80 percent of their fiber requirements. Renewable agricultural fiber crops can and will fill this void. It is already widely understood that one acre of industrial hemp can produce up to four times as much paper as an acre of trees. But although academia and industry know that industrial hemp is the superior fiber crop, hemp cannot be grown or effectively researched under existing rules.

And yet industrial hemp is not a new crop. Its cultivation and use date back thousands of years. Twenty-nine countries, including England, Germany, and Canada, have expanding hemp industries.

Meanwhile, American agriculture is experiencing significant price and income pressures, which are contributing to decline and decay in rural communities. Our agriculture needs new high-value crops and alternative crops to raise on a growing number of under-utilized acres and on land coming into production from previous farm program set-asides. To continue to expand the acreage of existing crops will only deepen the price and income crises. The increasing extent of hemp imports into this country and the deepening of American companies' investments in hemp operations overseas clearly demonstrate the need for America to reevaluate its policies governing this crop.

On the whole, America is a leader. On the issue of industrial hemp, we are a distant follower. And the question is very time-sensitive. If we are to build a hemp industry there is much ground to be made up. A growing number of our agricultural, manufacturing, academic, financial, and environmental leaders are urging that we move to com-mercialize this potentially valuable crop and the industrial base it can support.

The United States is an island of denial in a sea of acceptance. I am sure that you will find the facts in this book to be very enlightening.

Erwin A. Sholts, Economist

Director of Agricultural Development and Diversification
Wisconsin Department of Agriculture, Trade and Consumer Protection

Chairman, North American Industrial Hemp Council

PREFACE

I grew up in one of the most air-polluted cities in North America: Los Angeles. On smoggy days, my eyes burned and it often hurt to breath deeply. Fortunately, when I was just three my father Phil, who was quite the adventurer, bought a three-acre island in the breathtaking San Juan Islands chain in Washington state. Each summer, we children were blessed by the chance to play among the trees and tide pools, watching the whales and boats pass by. These two distinct worlds— of industrial pollution and of the wilds of nature— contrasted sharply as I entered and exited the San Juans at the beginnings and endings of my summers. I developed a strong sense that living in natural beauty was blissful, while living in a polluted environment was not. Later, at age twenty, I was alarmed when an unidentified truck driver dumped a load of nuclear waste in a creek nine miles from my home in Altadena, California. Angry at what I perceived as horrible injustices against people and the earth, I started cutting out newspaper articles on pollution and fastening them to walls all around my house. With the belief that environmental problems could be solved, I became fascinated with learning everything I could about natural and human-made systems. In my early twenties, I would often spend five or six hours a day reading trade magazines about water, farming, biology, natural foods, computers, business, finance, and retailing. While I enjoyed reading such magazines (along with the *Wall Street Journal*

—a daily habit picked up from my father when I was fourteen), I little knew that such training would be invaluable in my current efforts to create a billion-dollar hemp industry. I was also fortunate to have a great many teachers along the way. Peter Dukich taught me to value soil and compost; Julia Russell of the Los Angeles Eco Home showed me practical steps to eco-conscious living; and Bill Mollison demonstrated his worldwide Permaculture plan for incorporating sacred aboriginal design principles into how we farm, build, and work. In 1987, I founded Harmonious Technologies as a way to combine my passion for the earth with my marketing and entrepreneurial skills. Observing that our society was burying leaves and grass in landfills, I undertook to inform people that such valuable organic matter can be recycled. In 1991, I wrote *Backyard Composting: Your Complete Guide to Recycling Yard Clippings,* of which well over half a million copies have been sold to date. Harmonious Technologies has helped seven hundred North American communities develop cost-effective programs for the distribution of home-composting bins. My vision is that the simple act of composting one's yard trimmings and kitchen scraps is a doorway into a sustainable future (and creates free fertilizer, as well). In conjunction with Smith and Hawken, I've recently developed a new bin, the HOME COMPOSTER™, which incorporates in its design everything I've learned in my years of composting experience. Currently, it's made from recycled plastic; I'm looking forward to the day I'll be able to use hemp in its manufacture. One of the reasons I like hemp is that it grows in compost and its products can be recycled back into compost. I first learned about hemp from watching a video copy of the United States government's 1942 motivational film *Hemp for Victory.* In 1994, I launched HEMPTECH, the Industrial Hemp Information Network, to help reintroduce this economically important yet misunderstood and under-utilized resource. Although we live in an era of unprecedented environmental damage, I remain optimistic about our opportunity to create a vibrant future for our planet. I hope this book stimulates your own interest in hemp, and opens new avenues to you for living and working in a beautiful and healthful environment.

ACKNOWLEDGMENTS

The publishing team at Chelsea Green shares a strong commitment to producing beautiful and informative books on sustainable living. Everyone from founder Ian Baldwin and publisher Stephen Morris to the outstanding editorial team of Jim Schley, Nicole Cormen, Rachael Cohen, and Michael Potts helped to guide and "massage" my manuscript. I also want to acknowledge the great help of Sonia Nordenson in the areas of fine-tuning and research.

Given the busy schedule resulting from running two companies while writing *Hemp Horizons,* I owe much of the book's success to my dedicated and sometimes-a-bit-overworked HEMPTECH/Harmonious Technologies team members, Carolyne Stayton, Wendy Millstine, and Mercedes Terreza.

In completing this project, I have had the good fortune to collaborate with numerous friends and associates across the globe. Four individuals contributed to key sections of the book: Dave West (farming), Tom Ballanco (law), Dave Olson (Japanese history), and John Lupien (historical research). The entire board of the North American Industrial Hemp Council offered inspiration and much assistance, including Erwin A. Sholts of the Wisconsin Department of Agriculture, David Morris of the Institute for Local Self-Reliance, Andy Kerr of the Larch Company, Craig Crawford of EcoDesign Group, Ken Friedman of Hemp Industries Association, Jeffrey W. Gain of Blue Ridge Company, Andy Graves of the Kentucky Hemp Growers Coop Association, Geof Kime of Hempline, and Curtis Koster of International Paper.

Numerous others (I'm afraid I will omit a few) contributed to the book, including Joe Hickey of the Kentucky Hemp Growers Coop Association; Chris, Boyd, and Stacey Vancil of Oxford Hemp Exchange/Industrial Ag Innovations; David Watson, Robert Clarke, and David Pate of the International Hemp Association; Candy Penn of Hemp Industries Association; Michael Karus of the nova Institute; Gero Leson, formerly of the nova Institute and now of Consolidated Growers and Processors; Mary Kane, Petra Sperling, and Ellen Komp of HempWorld; John Howell and Malcolm MacKinnon of *Hemp Times*; Ian Low of Hemcore; Flores de Vries and Marcel Hendriks of HempFlax; Don Wirtshafter of the Ohio Hempery; the Kentucky Hemp Museum; Carl Stroml of Rohemp; Janos Berenji of the Institute of Field and Vegetable Crops; Christian Vogel of the Institute for Organic Agriculture (University of Vienna); Cyril Esnault of La Chanvrière de L'Aube; Cynthia Thielen of the Hawaii House of Representatives; Grant Steggles of Australian Hemp Products; David Frankel of Frankel Brothers Hemp Outfitters; Darrel Novak of All Points East; Neils Peter Flint of the Danish Hemp Society; Agi Orsi of Ortex; Jan Wilson; Eric Werbalowsky; and John McPartland.

Hemp Horizons

British Minister of State for Agriculture Michael Jacks in one of Hemcore's hemp fields near Essex, England, in 1995.

Welcome to the World of Hemp

CHAPTER ONE

I WAS DRIVING through steep, forested mountains in central Oregon on my way back to town, where I had inadvertently left the notebook that contained my manuscript for this book. As I backtracked along the highway, I saw again the many huge, gaping clearcuts with their resulting soil erosion and visual blight. I couldn't help thinking how the cumulative effect of deforestation silts the streambeds and raises water temperatures, ultimately destroying the delicately balanced habitat that such species as salmon and steelhead require.

Traveling along such highways always creates a sense of unease deep inside me. Within another hundred years, this lush region may go the way of the Sahara Forest, now known as the Sahara Desert.

I remembered last using the notebook at a phone booth outside a local diner. As I pulled up next to the booth, I saw my manuscript, but the phone was in use by a man in a black cap with a large gun visible on his hip. I approached and clearly read the printing on his cap: DEA, letters that I knew stood for Drug Enforcement Administration.

After a minute, the man hung up the phone, and I politely introduced myself as the president of HEMPTECH, explaining that I had left

3

my manuscript on industrial hemp in the phone booth. He looked at me askance, but I took the opportunity to tell him how useful fiber hemp is, and that thousands of products, from paper and building materials to textiles, food, and cosmetics, can be made from it.

The agent seemed firmly convinced that "hemp" is just another word for marijuana. I explained how industrial hemp is grown commercially in France, Germany, the United Kingdom, and numerous other countries around the world. He replied that he had been to Europe and that there was little interest in hemp there. I countered by telling him that the United Kingdom has been increasing its planting acreage by 50 percent each year.

A glazed look had come over the man's eyes. He was listening, yet not really hearing, because as a representative of the DEA, he had to remember not to confuse reality with official policy.

I went to my car and returned with a small block of fiberboard supplied by HempFlax of the Netherlands. Gesturing at the denuded hillsides around us, I pointed out to him that some corporations are actually manufacturing hemp fiberboard as a replacement for wood. Gravely, the DEA man said that the fiberboard looked like hash (compressed marijuana resin).

I gave him a copy of my earlier booklet *Industrial Hemp* and drove away contemplating the meaning of this chance encounter.

I had left my notebook in that phone booth about twelve hours ago. Had he waited there all night, hoping that I would return? Ruling out this unlikely possibility, I reflected on the odds of leaving my hemp manuscript by the phone that a DEA officer would soon use.

I've been working for several years to revive the commercial cultivation of hemp, and have had several such chance meetings with the

DEA. It almost seems like they've been on my trail . . . or have I been on theirs?

One year before the phone booth encounter, I attended a reception at Waste Management's corporate office in downtown Los Angeles, where my friend Gary Peterson introduced me to a DEA agent. As we discussed the hemp controversy, I pointed out that the chair next to this fellow was made from hemp fabric. I asked the agent whether he knew that the annual conference of the American Farm Bureau had endorsed efforts to research the growing and processing of industrial hemp. He said he had never heard of that, so I invited him to come to the next meeting of our trade association, the North American Industrial Hemp Council (NAIHC).

Despite numerous invitations, however, no DEA representative has ever attended an NAIHC meeting. It rather hampers a developing industry when the regulatory agency concerned—in this case, the DEA—refuses to acknowledge the legitimacy of those within the industry. And it is ironic that the DEA provides research permits to grow marijuana across the United States, yet refuses to do the same for industrial hemp, the nonnarcotic variety of the plant.

Here is the crux of the problem confronting the revival of the American hemp industry today:

- Although hemp is not illegal to grow, possess, or sell in this country, it is heavily regulated and its cultivation is discouraged.
- The United States legislators who crafted our first anti-marijuana laws took pains to differentiate hemp—a traditional, widely cultivated commercial crop—from marijuana.
- The DEA, which regulates hemp, has disregarded these distinctions, despite ample evidence that hemp and marijuana are not the same.
- A combination of regulatory fervor and bureaucratic indifference has resulted in misinterpretation of the law, making it extremely difficult to obtain permits for domestic hemp production (see chapter 3).

Yet there are compelling reasons to resume commercial hemp activity in the United States and worldwide, as this book will show. The more one learns about the phenomenal economic, environmental, and agricultural potential of industrial hemp, the more one sees the absurdity of the lack of rational discussion on the subject.

In 1995, I was given a unique opportunity to debate the retired Southeast regional director of the DEA, Wayne Roques, on National Public Radio. Mr. Roques took the familiar agency stand that there is no difference between hemp and marijuana. The radio announcer then lobbed a question my way: "So what *is* the difference between hemp and marijuana?"

I replied that hemp grown for fiber, whether by George Washington in 1790, by Kentucky growers in 1935, or by United Kingdom farmers in 1995, has never contained psychoactive qualities. If one were to roll leaves from an industrial hemp plant into cigarettes and smoke them, no euphoric effect would be experienced, even if a thousand hemp cigarettes were smoked. The potentially psychoactive chemical in hemp is delta-9 tetrahydrocannabinol (THC). Industrial hemp generally contains less than 1 percent THC, while its chemically distant cousin, marijuana, contains 3 to 15 percent or more.

Mr. Roques responded that marijuana of the 1960s had no more THC than industrial hemp has now. This is, of course, untrue. I felt like suggesting that he just ask anyone who would admit to having "inhaled" in the 1960s; she or he would indeed confirm having experienced a psychoactive effect back then.

If you smoke some industrial hemp, I told him, you'll only get a headache, and if you smoke more you'll get a bigger headache. I pointed out that dozens of countries, including France, the United Kingdom, and China, have laws permitting farmers to grow hemp while still prohibiting marijuana cultivation as a jailable offense. And I related how, the first time hemp fields were planted legally in modern England, some local teens chopped down a few plants, thinking they had found marijuana. Of course, after they had coughed a lot and gotten mild headaches, word spread around that hemp was to be avoided.

Former agent Roques now cited further misconceptions about hemp: Well, in any event, it is not economical; they actually get heavy subsidies in Europe for growing hemp.

Yes, I replied, they do subsidize hemp in Europe, because they see it as an important rotational crop in their move toward a more sustainable agricultural and economic system. In between the call-in questions from listeners, I took the opportunity to add that several Fortune 500 corporations such as International Paper are actively researching the use of hemp fiber for a variety of paper and building-product uses.

Later that day, it seemed my phone acted up a bit. But perhaps my paranoia was justifiable for someone who had just publicly debated the DEA on the subject of botanically distinct varieties of *Cannabis*. Should garden poppies be plowed under because another closely related member of the genus is grown for heroin? Yet this very logic is used to maintain the current suppression of hemp farming in the United States.

THE DIFFERENCE BETWEEN HEMP AND MARIJUANA

I have had the chance to develop many good friendships with advocates of industrial hemp around the world, and it seems that members of the media always call three or four of us when doing a piece. At these opportunities, it seems we hemp advocates always share our own particular favorite story or explanation.

Joseph W. Hickey, Sr., executive director of the Kentucky Hemp Growers Cooperative Association, likes to say: "Calling hemp and marijuana the same thing is like calling a rottweiler a poodle. They may both be dogs, but they just aren't the same."

Dr. William M. Pierce Jr., of the Department of Pharmacology and Toxicology at the University of Louisville School of Medicine in Kentucky, explains:

> It is absurd, in practical terms, to consider industrial hemp useful as a drug. . . . While a person could choose to use hemp in this way, it is unlikely that he or she would repeat the behavior, due to the unpleasant side effects. . . . It is possible to get drunk on "non-alcohol" beer, but no one does it. The amount necessary is far too great. Nutmeg contains a psychoactive substance that could be abused, but no one does it (too many side effects). In the areas that ban the sale of alcoholic beverages, it is easy to find mouthwashes that are 40 to 45 proof, and yet no one abuses them. In summary, the use of industrial hemp as a psychoactive substance is extremely unlikely, due to the large doses required and the side effects that would be encountered.

For those seeking a definitive comparison of hemp and marijuana, consider the following excerpt from the Canadian government's "Fact

Sheet on Regulations for the Commercial Cultivation of Industrial Hemp":

> Hemp usually refers to varieties of the *Cannabis sativa L.* plant that have a low content of delta-9 THC (tetrahydrocannabinol) and that are generally cultivated for fiber. Industrial hemp should not be confused with varieties of *Cannabis* with a high content of THC, which are referred to as marijuana.

While the leaves of hemp and marijuana look the same, one can easily tell from a distance that hemp is different from marijuana. Stands of fiber hemp are planted densely, at a rate of three to five hundred plants per square meter. Hemp plants are very tall, ranging in height from six to sixteen feet, with the majority of each plant comprised of thin stalks with no branches and relatively few leaves. In contrast, marijuana is planted one to two plants per square meter and is quite bushy, with lots of wide branching to promote flowers and buds. The distinction in the public mind between industrial hemp and psychoactive marijuana is key to the revival of a both proven and promising natural resource.

OLD PLANTS RETURN ANEW

Up until the late 1800s, humanity relied primarily on plant materials for making soaps, resins, paper, textiles, printing inks, building supplies, and so on. The chemical revolution of the twentieth century replaced many of these agricultural products with new synthetic ones. Today, chemical-dependent processes dominate the production of commodities, such as the pulping of trees to make paper and the intensive use of herbicides to raise cotton.

As the next millennium dawns, it is clear that our ever increasing demand for material goods is harming many societies and the Earth itself. For example, families have been forced from their forest homes in Indonesia so that the populations of industrialized nations can have cheaper paper and construction materials. Salmon fishermen in our own Pacific Northwest are losing their livelihoods, due in part to the destruction of forest watersheds. Nonsustainable logging practices and the long-term application of harsh farm chemicals have had numerous well-documented, negative effects on the health of waterways and soils.

We can learn to choose and use agricultural products in ways that will reduce such severe impacts. A return to regionally grown, sustainably harvested fibers is pivotal to conserving our forests, fisheries, and topsoils, while providing dignity to people who cannot afford paid lobbyists in Tokyo or the halls of Washington, D.C. Favoring such natural fibers as hemp; jute; flax; sisal; kenaf; ramie; bamboo; organic cotton; recycled fibers; and the residual straws of corn, wheat, and rice— instead of synthetic, chemical-dependent, or tree-dependent fibers— will help us to make a positive, Earth-friendly transition into the new millennium. Of course, wood is also a "natural fiber" that can be harvested in a sustainable manner to ensure healthy forest ecosystems (see chapter 7).

Organizations such as the United States Department of Agriculture's Alternative Agricultural Research and Commercialization Corporation (USDA AARC) are working to expand opportunities for farmers and entrepreneurs to create a rural economic-development renaissance by growing, processing, and marketing fibers. The AARC has leveraged private funds to enable small firms to transform kenaf into fine writing paper and to convert soybeans and recycled newspaper into marble-like tile. The return of industrial hemp is part of a greater trend that is transforming agricultural crops into new products that are both economically and environmentally sound.

AN INVALUABLE BIOLOGICAL RESOURCE

As will be further discussed in chapters 2 and 7, the world is slowly moving toward a carbohydrate economy (one that relies on plant materials) and away from a petroleum economy. Hemp fits well into this resource shift, and can transform our overreliance on petroleum-based products and processes.

Imagine a crop more versatile than the soybean, the cotton plant, and the Douglas fir combined—one whose products are interchangeable with those made from timber, cotton, or petroleum; one that grows like Jack's beanstalk with minimal tending. Industrial hemp *is* such a crop.

This hardy plant thrives in most climates, grows rapidly, and is highly resistant to disease and insects, largely eliminating the need for costly herbicides and insecticides. Its dense, fast-growing stalks discourage common weeds. In addition, hemp yields three to seven tons of dry fiber stalk per acre, depending on the variety and on the climate in which it is raised.

After hemp has been harvested, the field remains virtually weed-free for the next crop. In the twenty-nine countries allowed to cultivate hemp, this fact alone is saving farmers untold thousands of dollars while improving soil and water quality by eliminating the need for agricultural chemicals.

Another major benefit for farmers is hemp's soil-enhancing ability. Because this plant grows densely, reaching a height of several feet very quickly, the soils in hemp fields maintain cooler temperatures than do soils planted with other row crops. As the hemp stalk shoots up into the sky, the plant's leaves drop to the ground, mimicking a self-mulching forest ecosystem. Additionally, hemp's deep taproot helps aerate and break up compacted soils; in combination with cooler soil and mulch from falling leaves, this creates an ideal humus-rich growing medium.

The Canadian government recognizes the agronomic benefits of hemp, as summarized in Agriculture and Agri-Food Canada's December 16, 1994 *Bi-weekly Bulletin* (printed on Hemp paper). "In an age increasingly interested in sustainable agriculture and crop diversifica-

Hemp's dense planting smothers weeds, while falling leaves, rich in nitrogen, improve the soil as in a self-mulching forest ecosystem.

tion, hemp offers attractive possibilities. It is exceptionally disease- and herbivore-resistant, can be easily grown in a wide range of agricultural systems, and is an excellent rotation crop that eliminates weeds."

Ian Low, director of Hemcore, the United Kingdom's leading hemp-processing firm, concurs: "Many of our farmers have been pleasantly surprised to see increases of yield of 10 percent or more when planting winter wheat after a rotation of hemp."

With the American tobacco industry in decline, tobacco farmers are now strongly interested in the cultivation of hemp. The Kentucky Hemp Growers Cooperative Association includes approximately one hundred farmers (most of whom grow tobacco) who are working to reintroduce hemp as a viable cash crop in the Bluegrass State.

Yet, for all its impressive attributes, hemp needs to be viewed from a balanced, realistic perspective. Several obstacles face prospective hemp farmers in the United States—obstacles perhaps even greater than the current laws impeding its cultivation. Hemp is still an un-familiar crop for farmers, and most hemp markets are still in their infancy.

Despite its long agricultural history, hemp currently is like Rip Van Winkle, just reawakening from its sixty-year slumber. This makes the plant now, in effect, a new resource for manufacturers and processors; as with any new resource, it will require creative technologies.

The majority of hemp today is grown in China and Eastern Europe, and relies on plentiful labor rather than mechanized harvesting and processing equipment. Currently, there are significant technical challenges delaying the large-scale, cost-effective processing of hemp fiber. Numerous Dutch, German, British, Austrian, American, Canadian, and Australian researchers and entrepreneurs are working to solve such challenges. Yet, until the industry develops some state-of-the-art hemp-processing technologies, hemp will remain a minor niche crop, however great its promise.

POTENTIAL HEMP PRODUCTS

In October 1994, London's *Financial Times* described the resurgence of hemp in this way: ". . . fiber hemp . . . is making a comeback in Europe and the United States as an ecologically friendly raw material for clothing and paper."

Part of the reason for this comeback is the wonderful variety of raw materials derived from the hemp plant—six different materials in all: long bast fiber, medium fiber, short core fiber, seed, seed oil, and seed meal.

Along with hemp, such plants as flax, kenaf, jute, ramie, urena, and nettle are all sources of long bast fiber, which comprises the outer part of the stalk. Approximately 30 percent of hemp's tall, thin stalk is made up of this long bast fiber (historically referred to simply as "fiber"), and 70 percent is short core fiber (called "hurds" or "shives").

The process of "retting" (derived from the word "rotting") separates the hemp stalk's long bast fibers from the core. Approximately 75 percent of the material left over from the fiber-separation process is a medium-length fiber ("tow"). The other 25 percent (historically referred to as "long line") is the fiber of greatest length ("long fiber"). The hemp seed can be used whole or crushed into oil. The resulting pressed residue is often called "seed meal" or "seed cake."

(1) Top of male plant, in flower; (2) top of female plant, in fruit; (3) seedling; (4) leaflet from a large, eleven-parted leaf; (5) portion of a staminate inflorescence, with buds and mature male flower; (6) female flowers, with stigmas protruding from hairy bract; (7) fruit seed enclosed in hairy bract; (8) fruit, lateral view; (9) fruit, end view; (10) glandular hair with multi-cellular stalk; (11) glandular hair with short, one-celled invisible stalk; (12) non-glandular hair containing a cystouth.

Illustration by E. W. Smith from The Great Book of Hemp, *used with permission of Inner Traditions.*

Long fiber

- has long, strong strands—superior to cotton—that are very desirable for textiles;
- has antimildew and antimicrobial properties that are particularly useful for sails, tarps, awnings, and carpets;
- is biodegradable and serves as an environmentally sound substitute for fiberglass.

Medium fiber

- has low lignin (resinous plant glue) levels that make it ideal for paper and nonwoven applications;
- shares the bast fiber's antimildew and antimicrobial properties, so it is well suited for medical applications and hygiene products, such as diapers and sanitary napkins.

Short core fiber

- is up to twice as absorbent as wood shavings, making it an excellent choice for packaging and animal bedding;
- serves as a direct, often sturdier replacement for wood in construction materials;
- blends easily with lime to create a strong yet lightweight concrete or plaster;
- is biodegradable and serves as an environmentally responsible material for use in manufacturing plastics.

Seed

- is a highly nutritious protein source, better tasting and more digestible than the soybean;
- equals the soybean's versatility, and can be processed into milk, cheese, ice cream, and margarine, among other foods;
- is favored for making birdseed.

Seed oil

- has the highest volume, among edible oils, of essential fatty acids, so it makes a superb nutritional supplement;
- tastes better and has a longer shelf life than flax oil (another good supplement);
- has antimildew and antimicrobial properties that make an ideal base for soaps, shampoos, and detergents;
- blends easily with other substances to produce lubricants, paints, and printing inks.

Seed meal

- supplies high protein and nourishment in food for people and in animal feed;
- serves as a mild digestive bulking agent;
- can be blended with other grains into flours for many kinds of baking.

See the section on Hemp Product Versatility in chapter 5 for further information.

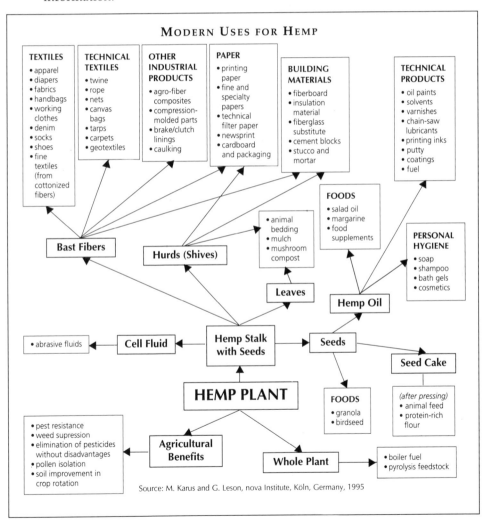

MODERN USES FOR HEMP

Source: M. Karus and G. Leson, nova Institute, Köln, Germany, 1995

SEPARATING THE OPPORTUNITIES FROM THE HYPE

Hemp offers a genuine and substantial boon to humankind, yet it is not a miracle crop that will save the world. Unfortunately, though, some of its more zealous proponents are prone to making wild, unsubstantiated claims that can undermine the credibility of the entire industry. The following are examples of a few such claims, with statements of the actual facts.

CLAIM: Hemp will eliminate the need to cut down forests.
FACT: Hemp can supplement our current fiber needs, but cannot replace trees unless we grow tens of millions of acres of hemp. Planting and harvesting new hemp fields on such a scale would require tremendous effort and energy. On the other hand, we could greatly reduce timber cutting for our paper and construction needs *right now* by using the vast quantities of agricultural straws (from wheat, corn, and rice) that typically are burned in the field.

CLAIM: Hemp oil can produce cheap fuel for automobiles.
FACT: Although soybeans are grown widely and yield several times more oil than does hemp, soy-oil–based diesel so far has been unable to gain even 0.1 percent of the fuel market. Further research combined with substantially higher gasoline prices may be required before there will be a widespread use of vegetable-oil fuels. Given the current political power of the oil industry, this is unlikely to come about in the near future. Inventor Rudolf Diesel designed his namesake engine to run on vegetable oils, and cars can and have run on hemp oil. Perhaps with the advent of the superefficient next-generation hypercars being developed by Colorado's Rocky Mountain Institute, fuels based on plants and/or vegetable oils may become more practical. For the time being, however, the cost of such fuels remains prohibitive—especially in the face of huge oil-industry subsidies.

CLAIM: Once hemp is grown in the United States, all hemp products will fall in price.
FACT: Until significant advancements are made in fiber processing, many hemp-fiber products will be much costlier than comparable cotton or timber products at current market prices. But the prices of hemp fiber, seed, and oil derivatives will come down substantially, due to a decrease in transportation costs.

CLAIM: Farmers will get rich growing hemp.
FACT: Hemp has never been a highly profitable crop for farmers.

Instead, it is a reliable rotation crop that provides weed control and improves soil conditions for the next crop. Furthermore, as food and farmland become scarcer worldwide, the value of grain crops will increase, and hemp will need to compete with such crops in order to return profits for farmers.

CLAIM: Industrial hemp varieties have no disease or insect problems. FACT: Hemp is subject to significantly fewer diseases and insect problems than are other fiber crops, including cotton, flax, and kenaf. When the crop is grown under sound agricultural management, with attention paid to soil fertility, hemp growers can usually avoid any use of pesticides. However, such growers are not immune to crop loss.

Those who separate the hemp hype from the hemp facts will be in the best position to evaluate and act upon the crop's numerous opportunities. Hemp can play a considerable role in rural economic development. Even prior to the crop's reinstatement in the United States, new jobs and businesses based on imported hemp fiber and seed are being created to produce goods for both local consumption and marketing to other regions. From a business perspective, hemp production can only increase once the plant has been reintroduced, because hemp is such an easy crop for farmers to grow. Astute hemp entrepreneurs have recognized the current bottleneck in fiber processing, and are instead focusing on products made from hemp seed and its oil. The processing and manufacturing of these products is far less capital-intensive than working with fiber, given today's available technology. Countries developing new seed and seed-oil businesses may realize a major benefit by stimulating the regional or domestic production of hemp seed. If hemp is to become a major crop once again, then American, Canadian, and Australian agriculture cannot continue to rely solely upon hemp seed imported from China and Eastern Europe.

THE RETURN OF INDUSTRIAL HEMP

A variety of powerful trends are molding and reshaping our society in ways that augur extremely well for hemp's return. In fact, many of these trends launch hemp from dusty dormancy to its rightful prominence as the fiber of the future.

The Digital Revolution

The Internet is allowing individuals and organizations to publish and communicate to millions of people around the world at very little cost. New ideas and products can expand like mushrooms popping up on the forest floor. Both Web sites and electronic mail have been used very effectively by hemp proponents and entrepreneurs.

Entrepreneurial Vision

Individuals from diverse locales and socioeconomic strata are starting their own businesses. Enterprises that can claim a well-thought-out business plan, access to modest amounts of capital, and a strong desire to succeed can bring about leading-edge changes in many industries. (See the business profiles at the end of each chapter and the resource list in the appendix.) Large corporations often follow the lead of successful entrepreneurs in their efforts to capture new markets. The success of a few pioneering hemp advocates in the fashion industry has encouraged such major companies as Adidas, Calvin Klein, and Giorgio Armani to make use of hemp textiles in their designs.

The Green Marketplace

Millions of people are now voting with their wallets by purchasing goods and services from companies committed to protecting our natural resources. For example, the natural-foods market is now a multibillion-dollar-per-year industry in the United States alone. Demand for Earth-friendly products made without harsh synthetic chemicals is also skyrocketing. The concept of holistic health care is transforming the conventional medical industry, as tens of millions of Americans now choose to include bodywork, acupuncture, herbs, and so on in their routine healthcare. Hemp is just beginning to penetrate this expanding green marketplace.

Fiber Wars

Pete Grogan, manager of market development for Weyerhaeuser Corporation's recycling unit, warned in a September 1994 *Wall Street Journal* article, "We're concerned there might be paper shortages this decade." The annual worldwide consumption of paper has risen from

14 million tons in 1913 to over 250 million tons in the 1990s. So much for the predicted paperless office! In the United States alone, our demand for wood products exceeds by weight our combined demand for all steel and plastic products. This is driving pulp and paper mills to obtain trees from sources farther and farther away from their factories, from the Amazon Basin to the Pacific Northwest to Siberia. As timber prices continue to rise, annual fiber crops such as industrial hemp, kenaf, and agricultural straws will become more economically competitive. Let's hope that hemp's return comes in time to spare the few precious old-growth forests that are still standing.

Biodiversity

Biodiversity describes the condition of a healthy natural system whose innumerable species of plants and animals interact with their habitats in a rich web of life. Our planetary biodiversity is in serious decline, due in large part to unsustainable logging practices and harsh, chemicalized agricultural and industrial methods that contribute to air and water pollution, soil erosion, and ever-dwindling forests. As we eliminate biodiversity, we disrupt delicate natural systems, reducing the supply of healthy food, air, and water. Yet, through education, awareness is growing that biologically diverse ecosystems serve as a synergistic glue for nature, and indeed for our planetary life-support system. Biodiversity must be protected if we are to survive. Since it is so well suited for organic, sustainable cultivation, industrial hemp has an important role to play in this effort.

Climatic Changes

The combined burning of huge quantities of petrochemicals and stripping of vast oxygen-releasing forests is accelerating climatic change around the globe. While many scientists may debate how fast this is occurring, very few will deny the fact that we face tremendous environmental challenges. Scientists quoted in the media as dismissing reports of the greenhouse effect as "bad science" usually turn out to be well-paid consultants to those industries responsible for the pollution. It has been scientifically documented that slight shifts in weather patterns can traumatically affect crop yields and forest health. Growers who have experienced erratic weather patterns over the past several years might do well to remember that hemp is much hardier than cotton, corn, or soybeans and can much better withstand a drought or a cold snap. Thus, farmers may want to raise some hemp each year as insurance against failure of a different crop due to inconsistent weather conditions.

Bioregionalism

Many communities are now realizing the benefits of producing goods locally versus relying on imported products shipped from great distances—products dependent on the burning of large quantities of petrochemicals for their transport. At present, this localized-production trend is more in the vision stage than in the actual implementation stage. Still, communities are beginning to see how an increase in their own production can benefit them economically, as dollars are recycled locally rather than whisked away to corporate headquarters in London, New York, or Tokyo.

The Rural Renaissance

All over the United States, people are leaving major urban areas in droves, to relocate to small towns or rural communities. The development of hemp processing and manufacturing centers in agricultural regions may help revitalize farming and provide jobs for those who adopt rural lifestyles.

The Simplicity Movement

The return to a simpler life, often referred to as the voluntary simplicity movement, is another strongly emerging trend. The best-selling book *Your Money or Your Life*, by Joe Dominguez and Vicki Robin, proposes earning and spending less as a positive antidote to today's stressful, credit-happy, overconsumptive lifestyle. Practitioners of simpler living, while buying fewer products, look for long-lasting, high-quality goods when they do purchase. Since hemp products can demonstrate both high quality and great durability, the hemp industry will surely benefit from the simplicity movement.

The Youth Culture

The youth culture of today will impact the business trends of tomorrow. By way of examples, recycling became popular at university campuses in the early 1970s. Alumni retained this new habit and later founded recycling programs across the country, often in direct opposition to status-quo institutional policies. Today, recycling is as American as apple pie. Hemp has the possibility to achieve similarly widespread acceptance in the coming decades. Americans aged fourteen to twenty-five years old are quite environmentally aware, and are very interested in hemp. Each week, I receive e-mail inquiries from students around the country who are looking for information to complete reports for school. They see hemp as a positive environmental solution, and view its current prohibition as an example of a corrupt political system.

Millennium 2000

As we approach the year 2000, there is a natural tendency to attempt to envision what the twenty-first century will offer. Some futurists suggest that innovative ideas will be welcomed and old, outmoded customs left behind. As a resource with a rich history and a potentially exciting future, hemp is well positioned for the new millennium: New jobs, a healthy environment, and everyday products that touch our skin (such as clothes and body care) or nourish us (such as food or nutritional supplements) all have personal immediacy and universal appeal.

HEMP GOES MAINSTREAM

Beginning in the early 1990s, after having slowed considerably during the previous twenty years, commercial hemp activity began to increase in the Western world. In 1993, the United Kingdom lifted its ban on hemp; the following year, Canada granted the agribusiness Hempline a research grant to grow the crop. By 1995, hundreds of entrepreneurs in Canada, the United Kingdom, Germany, Austria, the Netherlands, Australia, and the United States had started hemp companies.

In 1996, the North American Industrial Hemp Council (NAIHC) was founded by a coalition made up of environmentalists; entrepreneurs;

and representatives from industry, government, agriculture, and academia. As evidence of the strong commercial interest in hemp, representatives of Fortune 500 companies and of a state department of agriculture (Wisconsin's) were among the founders. This stamp of approval from government and big business has helped open many doors previously closed to industrial hemp.

By 1997, articles on hemp's commercial virtues were appearing regularly in such publications as the *Wall Street Journal*, the *Washington Post, US News and World Report*, and the *Kiplinger Letter*. Hemp has truly arrived in many segments of mainstream America.

THE HEMP HOUSE OF THE VERY NEAR FUTURE

While the following scenario may read like a futuristic fantasy, today's projections often foretell tomorrow's reality. After all, a hundred years ago, it would have seemed fanciful to think that so many household products one day would be made from wood composites and synthetic petroleum, as they now are.

So imagine that one day within the next ten years, you wake up in a house whose walls, roof, flooring, insulation, and paint are derived from hemp. You feel great after sleeping on your hemp-stuffed mattress, covered with soft linens spun from hemp fiber. Your feet sink into the hemp carpeting as you get out of bed and open the hemp drapes. It's a beautiful morning.

You jump into the shower, where you use soap, shampoo, and hair conditioner made from hemp. You step out onto the hemp bath mat, drying yourself with a superabsorbent hemp towel. You clean your ears with H-Tips (better than the old cotton swabs), and apply hemp-oil lotion, moisturizer, and lip balm. You make a mental note to buy some more hemp toilet paper, recalling how it wasn't too long ago that we were still cutting down centuries-old trees just to flush them away.

Opening your closet, you dress in hemp jeans, shirt, and jacket; put on hemp socks and shoes; tie the hemp laces; and grab your hemp wallet, which holds checks and currency printed on hemp paper.

You're hungry, so you walk into the kitchen with its hemp-based linoleum floor. You make some wheat-and-hemp-flour toast, and pour a glass of fresh, organic hemp milk. After eating, you make a salad with hemp-oil dressing to take to work. Then you wash your dishes, using hemp-oil dish soap and a hemp pot-scrubber, and put the dishes away in a cabinet built of hemp fiberboard. Sitting down on the hemp-framed and upholstered couch, you glance at a newspaper printed with hemp ink on hemp recycled paper, and learn that the hemp industry is now the largest agribusiness and the major job provider in your state. You turn on the stereo, which sits on a hemp fiberboard cabinet, and listen as music vibrates from speakers also made from hemp fiberboard. They contain specialty hemp paper for the speaker cones and are covered with black hempen cloth.

Leaving the house for work, you open the door of your car, built from strong, lightweight composites that include hemp. Relaxing into the driver's seat, luxuriously upholstered with hemp textiles, you rest your feet on floormats that look like rubber but are made from hemp. As you drive to your job at the new hemp-fiber processing facility, you pass farmers harvesting some of the locally grown hemp that is revitalizing your community's rural economy.

Hemcore

The United Kingdom's Hemcore is jointly owned by an agricultural supply company, Harlow Agricultural Merchants (HAM), and Robert Lukies, a farmer and seed processor from Essex, England. In 1993, Hemcore was instrumental in overturning the United Kingdom's ban on industrial hemp cultivation, and has pioneered many agricultural and commercial projects since then.

Ian Low, currently managing director of HAM and a director of Hemcore, reports that lack of excitement has never been a problem at Hemcore. The company has been engaged in developing apparel textiles, furnishing textiles, nonwoven textiles, paper pulp, livestock bedding, building materials, and other products. However, despite earning over £1 million ($1.6 million) in annual sales, Hemcore has encountered its share of obstacles. As Low explains:

> Whilst this crop is incredibly versatile and has many uses, we have found great problems in establishing clear routes to market in the different areas. Industry is naturally suspicious of agriculture, and when the crop concerned is *Cannabis sativa*, this suspicion takes a lot of overcoming. What is required is patient teamwork over a long period of time to prove what can be done, and when there are problems, to work through them. In summary, I am saying "lack of infrastructure," but now I believe this is starting to be overcome and the industry is beginning to establish credibility.

Among upcoming challenges, Low cites the following:

> Managing the huge expansion we believe will come, without losing the pioneering spirit that got it off the ground. We would also be concerned that the industrial hemp crop is not discredited or hijacked by the "legalize marihuana" lobby. Whilst we are well aware of the issues involved, we wish to see this excellent, environmentally sound crop achieve the importance it deserves as a farm crop and an industrial feedstock in its own right. . . .
>
> The drug issue is always there for us, as our farmers-growers have to renew their license every year with the [British] Home Office. We have had tremendous support from them and also from our Ministry of Agriculture, who want to see hemp established as a serious nonfood or annual renewable fiber crop. Any hint of a hidden agenda of campaigning for changes in the drug laws would cost us this valuable support.

I have mentioned the environment, but it must be stressed again. There are only two commercially viable fiber crops in a temperate maritime climate like ours; one is flax and the other is hemp, and environmentally hemp has by far the stronger case.

Hemcore
Station Road, Felsted, Great Dunmow
Essex CM6 3HL, England
Telephone: 441-371-820066
Fax: 441-371-820069
e-mail: stuart@hemcore.demon.co.uk

A Kentucky hemp field in the 1940s.

I am prepared to deliver . . . hemp in your port, watered
and prepared according to the Act of Parliament.

GEORGE WASHINGTON
from a 1765 letter

Hemp's Historical Prominence

CHAPTER TWO

F EW PEOPLE TODAY realize that the misunderstood hemp plant, *Cannabis sativa L.* of the family Cannabaceae, has played a vital role in world commerce for at least six thousand years. Though shifting social and economic trends have influenced the scope of its cultivation in recent centuries, hemp traditionally has been relied upon to supply humanity with a wide range of essential commodities.

THE ORIGINS OF CULTIVATED HEMP

Asia

China was the first region in all the world to cultivate and use hemp. The plant (*ma* in Chinese, as shown at right) was used for making rope and fishnets as early as 4500 B.C. The following was reported by Xiaozhai Lu and Robert C. Clarke in their article "The Cultivation and Use of Hemp in Ancient China," printed in the June, 1995, issue of the *Journal of the International Hemp Association*:

Hemp was one of the earliest crop plants of China. Through long-term efforts, the ancient Chinese domesticated hemp from a wild plant into a cultivated crop. . . . Ancient Chinese techniques of hemp sowing, cultivation, and processing developed rapidly and became fairly advanced. The earliest Neolithic farming communities along the Wei and Yellow rivers cultivated hemp along with millet, wheat, beans, and rice. The oldest Chinese agricultural treatise is the *Xia Xiao Zheng*, written circa the 16th century B.C., which names hemp as one of the main crops grown in ancient China (Yu 1987).

The article also quoted a long-ago writer, Ji Sheng:

Deep plow and fertilize the soil before sowing the seed. When spring comes, about February to March, select the dusk of a rainy day to sow seeds. . . . Fertilize the hemp with silkworm excrement when it has grown to one *chi* tall, and when it has grown to three *chi* tall, fertilize it with silkworm and pig excrement. Water the hemp frequently, and if there is much rain, the quantity of water should be decreased.

Chinese interest in the written word led to the use of hemp for making scrolls, which eventually brought about the development of the world's first paper industry. The Chinese art of paper-making reached Persia and Arabia in the eighth century. The writings of Confucius and Lao Tzu were transcribed on hemp paper (which lasts longer than paper derived from wood fibers), and thus the wisdom of these sages was handed down through time. Ancient China also cultivated the hemp plant in order to weave its fiber into cloth, and used its seed for food and oil.

Hemp spread beyond China in the third century B.C. The seed stock went to Korea, from which it crossed the narrow channel to Japan's southern island, Kyushu. A coastal Japanese cave painting depicts tall hemp plants, waves, horses, and strangely dressed people—perhaps the Korean traders who introduced the crop. The word "hemp" is expressed in written Japanese by the *kanji* character, signifying its Chinese origin.

As time went on, hemp successfully adapted to the Japanese climate and was well established by the third century A.D. It became the staple fiber for clothing as well as for specialized purposes, such as eel-fishing lines, *geta* (high wooden sandals) straps, and *washi* (fine paper). But perhaps hemp's most fascinating role evolved in Japanese spiritual life. In the rituals of Shinto, the indigenous religion, hemp symbolizes purity and fertility; the ancient shrine at Taimdo (literally, "hemp shrine") near Osaka is dedicated to this plant. In Shinto and Buddhist temples, certain symbolic objects—bell ropes, purification wands and curtains (*noren*), and priests' robes—are made of hemp. Zen scholars and warriors (*samurai*) have expressed hemp's inspiration in *haiku* (poetry), *aikido* (a martial art), and other traditional arts.

Trade and communication among Japan, China, and Korea faded over the next few centuries, although Japanese scholars still traveled to China to study science, medicine, and agriculture. They learned to administer hemp preparations for ailments such as constipation, asthma, skin problems, and poisonous bites, as well as for promoting general vigor and for worming animals.

Hemp's high resale value brought economic strength and power to feudal Japanese shoguns and kept humble farmers busy at labor-intensive production. The hemp leaf became a common motif in Japanese fabric, where it still appears often in modern quilts, kimonos, and *noren* curtains. Apart from silk for the wealthy, hemp remained Japan's primary clothing fiber until the seventeenth century, when cotton was introduced. The new plant's high yields (produced by heavy fertilizer use and mass-processing methods) lowered its cost and popularized this fiber among the growing urban working class. Hemp fabric became somewhat exclusive—reserved for special garments and the upper class.

Despite its diminished role in clothing, hemp continued its dominance as a raw material with many practical applications, through the nineteenth century. Rural Japanese blended hemp fiber with seaweed, broom straw, and other plants to make conical snow hats and packboards for transporting heavy loads over the mountainous terrain. And, as in Europe during this period, Japanese military power relied upon hempen ropes and sails for its expanding navy.

Europe

The cultivation of hemp spread from Asia to the Mediterranean along ancient trade routes. The Greek historian Herodotus extolled hemp's role in the manufacturing of fine textiles in his *Histories*:

> I must mention that hemp, a plant resembling flax but much coarser and taller, grows in Scythia. It grows wild as well as under cultivation, and the Thracians [inhabitants of the modern-day Balkans] make

clothes from it very like linen ones—indeed, one must have much experience in these matters to be able to distinguish between the two, and anyone who has never seen a piece of cloth made from hemp will suppose it to be of linen.

Hempen rope and fabric from circa 400 B.C. have been found near Stuttgart, Germany, and hemp continued to be cultivated in Central Europe through the centuries. In A.D. 1150, Moorish Spain used hemp to found the first paper mill in the West. By the sixteenth century, the art of paper-making was firmly established in Europe.

During the Renaissance, the French writer François Rabelais praised hemp in *The Histories of Gargantua and Pantagruel*:

> Without it, how could water be drawn from the well? How would scribes, copyists, secretaries, and writers do without it? Would not official documents and rent-rolls disappear? Would not the noble art of printing perish?

The very canvas on which Renaissance artists created their master-pieces took its name from the genus *Cannabis* (from which the fabric was originally made), and the oil in the paint was often derived from hemp seed.

From the sixteenth to the eighteenth centuries, hemp and flax dominated among the fiber crops of Asia, Europe, and North America. French, Dutch, Spanish, British, German, and Russian trading ships—including those that brought the first explorers and colonists to America—were rigged with ropes and sails made from hemp.

The British empire was founded on naval superiority, which in turn required vast quantities of hemp fiber to keep its vessels swift and sure. No wonder then that hemp was the most important crop in Britain's economy during the 1700s. Villages named Hempstead, Hempton, and Hampshire, especially common in southern England, attest to hemp's vital role. To fight a war without reliable access to high-quality hemp would have been as unthinkable as doing without pig iron for cannon shot. To assure the empire's hemp supply, the British monarchy made its cultivation compulsory, and looked to colonies in Australia and North America, as well as to foreign trading partners, to supplement domestic crops.

By the 1700s and 1800s, Russia's largest agricultural export was hemp, which supplied sails and rigging for American, Canadian, and

European ships. Meanwhile, Britain's arch-rival, France, had more than eight hundred thousand acres of hemp under cultivation. During wartime, hemp supplies were often cut off to punish enemy nations, so each country endeavored to maintain its own source of supply.

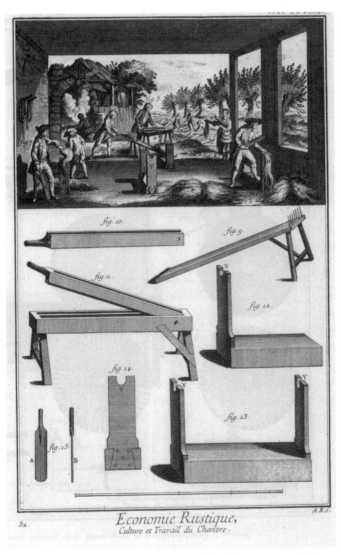

Eighteenth-century French tools for breaking hemp's strong fibers.

The New World

Factories for hempen rope and cloth developed not only in colonial New England and Canada but also in colonial Mexico and in Central and South America, through active trade with Spain. In Chile, hemp was cultivated with particular success to supply the ships of the Spanish conquistadors.

In New England, hemp was an indispensable raw material. Many of the American colonists' Bibles and maps were printed on hemp paper, and some of their lamp oil came from pressed hemp seeds. The high value placed on hemp fostered an early recycling industry, which converted old clothing, rags, ropes, and sails into paper. Hemp was so vital commercially that in 1640 the governor of Connecticut declared that every resident of the colony must grow the plant, to keep pace with the burgeoning need for marine cordage. In 1671, to stimulate hemp production and reduce imports, the Colony of Maryland offered local growers one pound of tobacco for every pound of hemp raised domestically. In 1700, Massachusetts required manufacturers of cordage to use fiber produced within the colony, and neighboring colonies followed suit.

America's founding fathers were strong supporters of hemp; in fact, both George Washington and Thomas Jefferson were long-time hemp farmers. The first two drafts of the Declaration of Independence were written on hemp paper.

Colonial women sewed hemp linens for the Continental Army, without which more soldiers might have frozen to death at Valley Forge. At the time of the Revolutionary War, the United States Navy's mightiest battleship, the U.S.S. *Constitution*, carried sixty tons of hempen rope and sails, including an anchor line that was two feet in diameter.

As surely as the twentieth century could be called the Age of Petroleum, the eighteenth century was the Age of Hemp.

A TEMPORARY DECLINE

In the first half of the nineteenth century, hemp maintained prominence in American industry. In the mid-1800s, more than 160 factories in Kentucky manufactured hemp bagging, bale rope, and cordage, employing several thousand workers. In regions where hemp was prominent, dozens of American towns still bear its name, including Hempfield, Pennsylvania; Hemphill, Kentucky; Hemp Island, Florida; Hemphill Bend, Alabama; Hempstead, New York; Hemp, Georgia; Hempton Lake, Wisconsin; Hempfield Lake, Mississippi; and Hempfork, Virginia.

By the end of the nineteenth century, however, hemp's vital role in commerce was in global decline. The advent of the steam and petroleum engines had severely reduced demand for hempen rope and sails, while the invention of the cotton gin had greatly cut labor costs in the cotton industry.

The lack of a mechanized method for hemp harvesting and processing severely limited hemp's economic competitiveness. Much of the remaining demand for hemp cordage in Canada and the United

Kentucky field hands loading raw hemp stalks.

Hemp in the Bluegrass State

The fertile soils of Kentucky's famous Bluegrass region made that state the leading hemp producer in the United States, a fact in which Kentuckians took great pride. *A History of the Hemp Industry in Kentucky* by James J. Hopkins, a book published in 1951, details the lore of hemp in the Bluegrass State. Kentucky hemp was world-renowned for its height, hardiness, fiber yield and quality, and rapid maturation. One of the challenges facing Kentucky hemp farmers and manufacturers was the labor-intensive process of breaking down the rigid hemp stalks. In earlier years, much of this hard work had been done by slaves. With the abolition of slavery after the Civil War, and the continuing expansion of the cotton industry, hemp acreage in Kentucky declined considerably toward the latter part of the nineteenth century.

Fortunately, Kentucky developed a strong interest in and reputation for hemp-seed production. In the early 1900s, through the work of the United States Department of Agriculture (USDA) agronomist Dr. Lysterr Dewey, an extensive breeding program helped to further refine Kentucky hemp.

States was supplied by Russian imports, which were cheaper because of the low wages paid to Russian serfs. Inferior-quality baling twine began to be manufactured from jute, sisal, and Manila hemp, which actually is not hemp at all.

The 1892 World's Fair, held in Chicago, featured hundreds of architectural columns made of artificial marble (plaster of Paris combined with hemp). But by the early 1900s, hemp was in general use only for cordage and for specialty seed and oil (the seed was used in bird feed, and the oil was an ingredient in paints and varnishes).

Gathering retted hemp stalks to be placed in vertical hemp shocks.

Nevertheless, there was still a flicker of interest in the possibility of expanded applications for hemp, particularly as a raw material for paper. The 1904 edition of Canada's *Pulp and Paper* magazine published several studies, including "Hemp Waste for Paper," that reported favorably on these prospects. A 1910 United States Department of Agriculture (USDA) article, "Utilization of Crop Plants in Paper Making," stated:

> Evidence has been gathered that certain crops can probably be grown at a profit to both the grower and the manufacturer solely for paper-making purposes. One of the most promising of these is hemp.

With the outbreak of World War I in 1914 and the resulting drop in hemp imports from Russia and Italy, hemp acreage increased within the United States as American manufacturers looked to domestic sources for fiber. The 1917 *Yearbook of the United States Department of Agriculture* reported: "The total acreage [of hemp] of the entire country doubled annually, reaching an estimated height of forty-two thousand acres in 1917."

Canadian hemp pioneer Howard Fraleigh (left) and colleagues stand proudly in front of their hemp field, circa 1930s.

Andrew Wright, agronomist with the University of Wisconsin Agricultural Experiment Station and a leading hemp researcher, wrote in the August 1918 *Country Gentleman*:

> From a position of insignificance, hemp has become within the past few years a crop of national importance—second only to cotton as the greatest fiber crop of the United States. . . . Thread for army shoes, twine for the grain harvest, caulking for our ships—surely hemp should be reckoned among our foremost war crops.

A TECHNOLOGICAL BREAKTHROUGH

After World War I, fiber importation resumed and the demand for domestic hemp production subsided. But the war had occasioned a variety of experiments concerning the separation of hemp fibers. At the time, only about 15 percent of the hemp stalk—the part containing the longest fibers—was usable. (Since then, selective breeding has increased the long bast portion of the stalk to about 30 percent.) The remaining shorter core fibers, or hurds, were burned in the fields as waste.

In 1916, a major breakthrough for the hemp industry was heralded by the publication of USDA Bulletin No. 404: *Hemp Hurds as Paper Making Material.* The bulletin, which was printed on hemp-hurd paper, announced a prototype decorticating (fiber-separating) machine that would greatly reduce hemp's labor costs, improve paper quality, and conserve forests by providing a low-cost, abundant source of pulp to fill the world's growing need for paper. It further stated:

> Every tract of ten thousand acres devoted to hemp raising year by year is equivalent to a sustained pulp-producing capacity of forty thousand acres of average pulp-wood [timber] lands. In other words, in order to secure additional raw materials for the production of twenty-five tons of fiber per day, there exists the possibility of utilizing the agricultural waste [the core fiber previously burned in the fields] already produced on ten thousand acres of hemp lands instead of securing, holding, reforesting, and protecting forty thousand acres of pulp-wood lands. . . . After several trials, . . . paper was produced that . . . would be classified a Number One machine-finishing paper.

A hemp hand brake   *A hemp machine brake*

Hemp had now received U.S. government sanction as a viable and important cash crop, one that could replace forest products as a source of paper pulp. It remained only for someone to manufacture and distribute the much-needed fiber-separating machinery.

Hundreds of inventors had tried to come up with an efficient process for capturing all of the hemp plant's useful fiber. In 1917, inventor George W. Schlichten presented his decorticating machine, which economically separated hemp fiber from the core, and for which he was issued a patent. This new technology reduced labor costs dramatically, and created the possibility of the American hemp industry assuming a stronger economic position than ever before: The stalk was now up to 95 percent usable, a threefold improvement over previous yields. Yet the paper trail of Schlichten's invention mysteriously disappears in the mid-1920s. During the following decade, several companies built factories that used innovative fiber-separating equipment. Was their technology based on Schlichten's invention? We may never know.

WOOD PULP VERSUS FARM "WASTE"

As American demand for fiber increased during the 1920s there was tremendous concern about deforestation and about the underutilization of agricultural residues that could economically substitute for wood pulp in the making of paper and building products. Many farmers, researchers, businesspeople, and politicians advocated using the agricultural "waste" from hemp, corn, wheat, flax, cotton, and other crops.

On January 25, 1927, Congressman Cyrenus Cole of Iowa introduced legislation for a $50,000 appropriation so that the Bureau of Standards could research the utilization of farm "waste" on a commercial basis:

> Can we not make something out of these wastes? . . . I believe the
> farm must be coupled with the factory and the factory with the farm.
> . . . We must find more uses for our so-called raw products.

While Congress approved Cole's appropriation bill, the USDA was overtly hostile toward it, on the grounds that the proposed study duplicated investigations already being conducted by the agency Bureau of Forestry and Forest Products. This is one of the first recorded instances of a blatant institutional bias in favor of forest products over agricultural crops.

In March 1929, the *New York Times* ran an article prominently headlined "Cornstalk Paper Not Satisfactory," describing cornstalks as a poor substitute for wood-pulp newsprint. The *Times* piece was released just eight days after Senator Thomas Schall of Minnesota had delivered a speech before the Senate in support of alternative farm sources for paper. A few weeks later, another negative article appeared, this one written by the head of the Newsprint Institute.

During this period, the species Southern pine was being explored as a new source of wood pulp. The idea received a positive response both from the media and in Congress, in marked contrast to the negative reception toward Cole's and other legislation promoting the use of farm residues. In 1932, the USDA released an influential report supporting the development of Southern pine, while stating that hemp was unsuitable for the production of paper—quite a turnabout from 1916's Bulletin No. 404!

To understand this renewed emphasis on wood pulp over farm "waste" in the paper industry, it is important to bear in mind that hemp and other agricultural residues require significantly less chem-

---

Homegrown Raw Materials

Why use up the forests which were centuries in the making and the mines which required ages to lay down, if we can get the equivalent of forests and mineral products in the annual growth of the fields? I know from experience that many of the raw materials of industry which are today stripped from the forests and the mines can be obtained from annual crops grown on the farms. . . . Industrialization of crops will also have the advantage of making a considerable saving to the manufacturer who learns how to accomplish it. . . . The best possible working plan for any man in our civilization is to have one foot on the soil and the other in industry.   —Henry Ford

---

ical treatment and less energy for conversion to paper, as compared to what is needed to break down tree fibers. DuPont and other large chemical manufacturers controlled the patents and processes for making paper from tree pulp. It is hard to prove whether these powerful vested interests influenced the choice of wood products over farm products. In any event, it happened, and denuded hillsides and downstream pollution from wood-pulp mills are now worldwide environmental concerns.

THE DECLINE OF THE FAMILY FARM

Many farm historians describe the 1920s as the beginning of a regulative assault on the American family farm, during which the bedrock of the agrarian democracy envisioned by George Washington and Thomas Jefferson started to slip away. At that time, many farmers maintained self-sufficiency by converting part of the corn they raised into alcohol fuels for their vehicles and farm equipment. With the

Hemp stalks ready for shipment to the mill.

Prohibition of 1919, the government was given the power to destroy corn-alcohol distilleries, forcing the farmers into dependency upon gasoline and diesel.

Private banking interests took control of the monetary system with the formation of the Federal Reserve in 1913 and, several years later, the Internal Revenue Service to collect income tax. This form of taxation was arguably unconstitutional because it was never ratified by the required three-fifths of the individual states.

THE 1930S: ANOTHER EXPANSION

As the 1930s began, the annual harvest of American hemp occupied only a few thousand acres. This minor industry relied on the traditional labor-intensive processing of hemp fiber into rope and twine. But as the Great Depression wore on, American farmers and their communities desperately sought new crops and business opportunities for rural America.

Hemp proponents began to promote a new industry based upon innovative fiber-separating technologies such as Schlichten's decorticating machine. This fledgling industry was based upon the science of chemurgy (a term coined by Dow Chemical biochemist William Hale), which sought to combine agriculture and organic chemistry. Founders of the chemurgy movement included Henry Ford, Thomas Edison, and George Washington Carver, who shared a dream of seeing farm products replace timber and imported oil as sources for fibers, plastic, fuels, and lubricants. The chemurgists operated on the premise that "anything that can be made from a hydrocarbon can be made from a carbohydrate."

The new hemp industry was predicated upon a chemurgic vision of cellulose as a golden source of economic salvation, supplying the raw material for an array of such products as hemp paper and plastics. This vision was in distinct contrast to the state of the archaic hemp industry, which now produced only minor quantities of rope and textiles.

By 1937, industrial hemp companies had sprung up across the Midwest, which now eclipsed Kentucky and Missouri as the nation's hemp capital. Important firms in Illinois included the Amhempco Corporation, Fibrous Industries, and the World Fiber Corporation in Chicago; the Illinois Hemp Company in Moline; and the Chempesco Central

Fiber Corporation in Champagne. Minnesota was home to Chempco and Cannabis, Inc., in Winona, the Hemlax Fiber Company in Sacred Heart, Central Fiber Corporation in Blue Earth, and Hemp Chemicals Corporation and the Northwestern Hemp Corporation in Mankato. Wisconsin had the Atlas Hemp Mills in Juno, the Badger Fiber Company in Beaver Dam, and the Matt Rens Hemp Company in Brandon.

The rise of the new Midwestern manufacturers substantially increased the amount of acreage planted to hemp. Between 1934 and 1937, the Northwestern Hemp Corporation, the Amhempco Company, and Chempesco each planted thousands of acres annually for specific cellulose-based products such as paper and plastics.

A 1938 *Popular Mechanics* article, "New Billion-Dollar Crop," proclaimed:

> American farmers are promised a new cash crop. . . . A machine has
> been invented that solves a problem more than six thousand years old
> . . . designed for removing the fiber from the rest of the stalk. . . .
> Hemp is the standard fiber of the world. It has great tensile strength
> and durability. It is used to produce more than five thousand textile
> products ranging from rope to fine laces, and the woody "hurds" . . .
> can be used to produce more than twenty-five thousand products,
> ranging from dynamite to cellophane. It can be grown in any state
> of the Union.

Certainly hemp cultivation was on the rise in the United States. But hemp's resurgence would be curtailed by the demonization of its disreputable cousin, marijuana, which was outlawed in 1937 by the Marihuana Tax Act.* With the passage of the tax act and its resultant red tape, the growing of hemp and production of its derivatives fell considerably (See chapter 3). Although the tax act permitted state or federal government-licensed hemp farming, obtaining such licensing became so difficult that most hemp farmers and processors simply gave up.

---

*This book uses the spelling "marijuana" throughout except as here, in quoted matter, and in historical contexts, where "marihuana" is used.

La Chanvrière de l'Aube

La Chanvrière de l'Aube is a well-established French farmers' cooperative specializing in the production of animal bedding and construction and insulation materials, but also supplying hemp fiber to manufacturers of paper, textiles, and fiber composites, as well as seeds and oil to makers of cosmetics and food products. Cyril Esnault described the company's history, speaking from La Chanvrière's offices in the middle of the vineyards of Champagne, France:

> Industrial hemp has been grown for years in France, especially in the Champagne area. It was used mainly for the paper industry. By 1973, European hemp seemed likely to disappear, so our farmers rallied together to create a cooperative for fiber plants, among which *Cannabis sativa* was a major one. From that time, La Chanvrière de l'Aube has kept on evolving. . . . Everything was different twenty years ago. At that time in Europe, we were perhaps the only ones to believe in industrial hemp. The transition from old to modern times was difficult to manage. Now the hard days are far behind us, but we keep them in mind.

La Chanvrière de l'Aube

La Chanvrière de l'Aube's round hemp bales heading to French factories.

La Chanvrière de l'Aube
Rue du Général de Gaulle
10200 Bar Sur Aube, France
Telephone: (33) 03 25 92 31 92
Fax: (33) 03 25 27 35 48
www.marisy.fr/chanvrier

Hemp stalks shoot for the sky in Canada, one of twenty-nine countries currently profiting from hemp fiber's resurgence.

I would say they [legitimate hemp farmers]
are not only amply protected under this Act, but
they can go ahead and raise hemp just as
they have always done it.

HARRY J. ANSLINGER, COMMISSIONER
FEDERAL BUREAU OF NARCOTICS

The Law and Politics of Hemp

CHAPTER THREE

HEMP'S LEGAL and political status in the United States, as in other parts of the world, is complicated by many factors. In the United States the cultivation or even possession of marijuana is a felony in some states and illegal in all fifty, with severe penalties enforced. Yet no federal law exists—or has ever existed—to prohibit the cultivation of industrial hemp in this country. Indeed, commercial hemp farming continued at various levels until the late 1950s—twenty years after Congress first outlawed marijuana in 1937.

Under current law, there is no reason why legitimate hemp farming cannot be resumed today. However, anyone cultivating hemp at the present time risks criminal prosecution.

If you find this state of affairs confusing, you are not alone. The confusion about the legal status of hemp hinges on the cumulative misinterpretation of laws aimed against marijuana. The legislators who originally crafted federal antimarijuana laws never intended to restrict hemp, but regulatory practice blurred the distinction between hemp and marijuana. Hemp growers and processors faced overwhelming red tape, heavy taxes, and the threat of arrest; in essence, the industry was criminalized, and it therefore declined precipitously. Today, this state of affairs could be reversed—with no need to legalize marijuana—if the distinction based on scientifically measurable THC content were upheld in the enforcement of existing laws.

CANNABIS: BOTANICALLY DIFFERENT VARIETIES

As illustrated in chapter 1, hemp and marijuana are two distinct varieties of the genus *Cannabis*. Just as much of the American public is underinformed about the differences between hemp and marijuana, so too is much of the American judicial and law enforcement establishment. The latter plant contains substantial percentages (3 percent or more) of the chemical tetrahydrocannabinol (THC), the psychoactive that produces the euphoric state that users call a high. The other variety, industrial hemp, contains negligible amounts of THC (less than 1 percent). These facts are scientifically verifiable today through the use of gas chromatography. Smoking hemp (if anyone did it) would cause a headache rather than a high, due to its high cannabidiol (CBD) and low THC content.

The contemporary Canadian researcher Ernest Small has written a two-volume study, *The Species Problem in Cannabis*, based on his extensive exploration of the subject. He summarizes the distinction as follows:

> This genus, through domestication, has been subjected to intensive disruptive selection, which has produced two kinds of plant. On the one hand, plants have been domesticated for the valuable phloem fibers in the bast. To maximize the quality and obtainability of these fibers, man has selected plants that are tall and relatively unbranched, with long internodes and with a relatively hollow stem. [The French naturalist] Lamarck termed such plants *Cannabis sativa*. Such domesticated plants have been characteristically grown in Europe, northern Asia, and North America.
>
> In contrast, man has also selected *Cannabis* plants for the ability to produce an inebriant. *Cannabis* synthesizes a resin in epidermal glands, which are abundant on the leaves and flowering parts of the plant. This resin comprises a class of terpenoid chemicals called the cannabinoids. Two are of particular importance: the nonintoxicant cannabidiol (CBD) and the highly intoxicant 9-tetrahydrocannabinol (THC). . . . Predominance of CBD characterizes the resin of fiber strains, and also of strains selected for the valuable oil content of the fruits

[seeds]. Predominance of THC characterizes "narcotic" strains of *Cannabis*.

Drug strains do not exhibit features related to harvesting the fiber. They are often fairly short, possess short internodes, are highly branched, and have comparatively woody stems. It was this type of plant that Lamarck named *Cannibis indica*. Such plants are characteristic of southern Asia and Africa, where *Cannabis* has been used for millennia as a source of the drug.

Much of the confusion between hemp and marijuana today can be traced to the antidrug campaigns that began in the 1920s. Even Matt Rens, the Wisconsin mill owner known as "America's hemp king" because of his contributions to the advancement of the industry, was, in his ignorance of the facts, influenced by these campaigns. In 1940, Rens was quoted in the *Milwaukee Journal* as saying, "We have enough marihuana on hand in stacks and in our warehouses to drug the nation, but I can't recall a single case since I've been in the business where farmers or help around here smoked marihuana or put it to improper use." What Rens had on hand in such abundance was hemp, not marijuana—enough hemp to cause widespread headaches, but hardly a nationwide high.

An examination of the *Congressional Record* chronicling the debate over drug laws in 1937 and again in 1945–46 shows the extent of misinformation among our elected representatives, and the confusion between hemp and marijuana continues to this day. Unfortunately, our nation's drug-enforcement policy is squarely based upon this fundamental inaccuracy. The policy has eradicated what was once considered a vital commercial crop, a form of agriculture employing thousands of Americans.

To set the matter in perspective, recall from the previous chapter that hemp farming was an important industry in America long before the English colonists considered seeking their independence, and hemp continued to provide essential raw materials for the young nation. Throughout American history, hemp farmers met with varying degrees of success and failure, boom and bust, depending upon weather conditions and market forces. The challenges of hemp farmers were much like those of any other American farmer—until 1937.

THE DEMONIZATION OF HEMP

The advent of Prohibition in 1919 was only one indicator of a mounting public sentiment against societal vices. The 1930s brought the first occurrence of films such as *Reefer Madness*, which demonized marijuana and thus tarred industrial-grade hemp with the same brush. Also in that decade, warnings bearing headlines such as "Killer Weed, Marihuana, the Greatest Menace to Society Ever Known" began to appear in newspapers.

In 1930, the Federal Bureau of Narcotics (FBN), precursor to today's Drug Enforcement Agency (DEA), was formed under the direction of the U.S. Treasury Department. Secretary of the Treasury Andrew Mellon appointed his son-in-law, Harry J. Anslinger, as commissioner of the new bureau. (It is worth noting that Mellon was then the principal banker to the largest emerging petrochemical firms, including DuPont, as well as to other corporations controlling vast timber acreage.)

In 1937, the increasing political antidrug sentiment brought the Marihuana Tax Act before Congress. Unfortunately for the hemp industry, no one had yet identified tetrahydrocannabinol as the psychoactive ingredient in marijuana. Still, it was known in Congress that there were two types of *Cannabis*: the kind that was smoked for its euphoric effects, and the important farm crop that was harvested for its fiber and seed. In fact, testimony that year before the Senate Committee on Finance and on the floor of both houses of Congress demonstrated pointed concern that the Marihuana Tax Act should control traffic in illegal drugs but not interfere with the legitimate hemp industry.

The pending legislation made the cultivation of marijuana a felony. Unfortunately for the hemp industry, most of the leaves were stripped off during harvesting, yet there were always some that remained attached to the stalks, creating a perpetual risk to hemp farmers and processors. If any leaves remained, the FBN had the right to confiscate, withhold, or destroy the hemp in question. The FBN considered these residual leaves no different than marijuana, which was now contraband.

As it continued to promote the passage of the Marihuana Tax Act, the FBN was well aware of the hemp industry's concerns. In late 1936, the agency learned of an experimental hemp crop being cultivated under the supervision of the *Chicago Tribune*, which reported on the progress of these experiments in a running feature titled "Day-by-Day Story of the Experimental Farms."

In a letter dated September 28, 1936, Commissioner Anslinger requested District Supervisor Elizabeth Bass, of the Bureau's Chicago office, to investigate these experiments. Bass visited the experimental farm and met with the *Tribune's* project director, Frank Ridgeway, who explained that the purpose of the project was to develop capabilities "towards the manufacturing uses of the fiber, pith, etc., and to commercial products." Ridgeway also expressed his lack of knowledge about marijuana, adding that he didn't know that hemp was the same plant.

Anslinger then requested further information from Bass on the demand for the machine used to harvest the "marihuana" (as he consistently referred to hemp); the regions of the U. S. where there was such a demand; and the uses of marijuana in those areas. Bass reported back in November:

> Objections raised by the manufacturing druggists who have slight need of the extracts of the *Cannabis* in medicinal compounds will be trifling when compared with the country-wide protests that will be raised as with one voice by the experimental stations everywhere developing the use of the fibers of the *Cannabis* plant stems for every variety of textile.

On June 12, 1937, J. P. Smith, an attorney from Mankato, Minnesota, contacted his congressman, Elmer J. Ryan, to voice concern about the Marihuana Tax Act pending in committee in the House. Smith was associated with the National Citizens Bank of Mankato, the primary financial backer of several hemp projects in southern Minnesota. He informed Congressman Ryan of the activities of the Farm Chemurgic Council, and mentioned that a commercial enterprise was interested in using hemp hurds as a raw-material source for plastic. Smith emphasized that a leading paper manufacturer had speculated that hemp in the Midwest was likely to develop into "a more important cash crop than soybeans." Smith further stated that he had heard of a variety of hemp which could produce a narcotic drug, but he did not believe that hemp raised for fiber purposes was the same variety. He summarized:

> We are unable to understand why such a bill should be proposed because, according to our information, it could serve no good purpose

and would embarrass, if not kill, an important agricultural development. . . . No one farmer raises any considerable acreage, the profits
are not large, and I do not believe any independent Minnesota farmer
would care to raise any crop under the license and direction of the
Federal Bureau [of Narcotics].

Less than three weeks later, Commissioner Anslinger replied to Congressman Ryan, who had passed Smith's letter on to the FBN. The commissioner explained that Smith had been misinformed regarding the
narcotic properties of hemp: All varieties contained a psychoactive substance. But Anslinger concluded his reply with an assurance that the
pending bill would not interfere with the legitimate hemp industry.

The escalating war against drugs had now introduced the confusion
of hemp with marijuana into public discourse, as evidenced by the
recorded comments of politicians and officials debating the new legislation. Even while ostensibly intended to protect the industrial hemp
industry, their statements only added to the general lack of clarity on
the subject. In fact, chemists for the Treasury Department charged
with identifying the psychoactive ingredient found in marijuana had
not done so by the time of the congressional hearings preceding the
tax act vote. Consulting chemist H. J. Wollner testified:

> The problem is not yet resolved. We are not yet in a position to know
> exactly what it is we are looking for, and, within four walls, I am per-
> fectly frank to admit that all the chemists I have met, who are inter-
> ested in this field, are at a complete loss when asked to prophesy the
> character of the narcotic principle which we are going to eventually
> disclose. The situation is as bad in the chemical literature as it is in all
> of the other phases. I should certainly be within reasonable bounds of
> correctness when I guess that 90 percent of the stuff that has been
> written on the chemical end of *Cannabis* is absolutely wrong, and, of
> the other 10 percent, at least two-thirds of it is of no consequence.

During the tax act hearings, Commissioner Anslinger cited stories
of brutal murders and other violent crimes in which marijuana had
been blamed for the defendants' actions. Many authorities took issue
with these claims, including Dr. William Woodward, legislative counsel for the American Medical Association, which testified against the
bill, stressing the importance of marijuana for pharmaceutical use.
The National Seed Oil Institute also lobbied strongly against passage,
pleading that hemp oil was essential to the production of paints and in
other industrial processes.

Clinton Hester, assistant general counsel for the Treasury Department, testified in favor of the Marihuana Tax Act before the Senate Committee on Finance. Discussing the possible impact on legitimate hemp farming, he stated: "The plant also has many industrial uses.... The production and sale of hemp and its products for industrial purposes will not be adversely affected by this bill."

Members of Congress urged their colleagues to pass the bill. Some, such as Congressman Robsion of Kentucky, continued to express concern about its possible negative effects on the hemp industry. Speaking on the House floor, Robsion inquired,

> I am opposed to the use of the drug taken from the hemp, but is this bill so drawn that it will not interfere with or injure the production of hemp for commercial purposes in a legitimate way?

His counterpart from California, Buck, replied, "This bill defines marihuana so that every legitimate use of hemp is protected."

Despite all objections, Congress passed the Marihuana Tax Act by a wide margin, and it was signed into law by President Franklin D. Roosevelt on August 2, 1937.

At this point, I must acknowledge that many hemp historians and researchers have put forth conspiracy theories regarding the Marihuana Tax Act and its intended effects upon the legitimate hemp industry. Of these, the most widely mentioned are the following:

- The William Randolph Hearst theory, which alleges that the newspaper baron, who owned vast timberlands, conspired against hemp with the DuPont Corporation, which owned patents on nylon and plastic, in order to remove a potentially formidable competitor;
- The Racism theory, which alleges that the antimarijuana campaign was not directed against hemp but targeted African-American jazz musicians and Hispanics in Southern border towns, who were sterotypically associated with marijuana use ("Marijuana" derives from the Spanish version of the name "Mary Jane");
- The Prohibition Bureaucracy theory, which alleges that government agencies formerly responsible for enforcing the alcohol ban had nothing against hemp but needed to create a new villain (marijuana) in order to justify their continuing existence after Prohibition was repealed in 1933.

A 1954 tax stamp for a U.S. farmer growing hemp.

There is no clear evidence that proves beyond doubt that the tax act was intentionally concocted to destroy the nonnarcotic hemp industry, or to create a climate that would favor the timber and petrochemical industries, although these outcomes indeed happened. On the other hand, from historical records, we are able to trace how government actions severely hampered the development of the hemp-fiber and seed-oil industries.

IMPEDIMENTS TO THE LEGITIMATE HEMP INDUSTRY

Despite assurances to the contrary, the passage of the Marihuana Tax Act in effect crippled industrial hemp. Commercial hemp farming in America continued. However, agents of the Federal Bureau of Narcotics, armed with the new statute, began to clamp down on the industry, especially in Iowa, Illinois, and Minnesota—the areas where hemp farmers and millers were using new cellulose technologies to develop innovative hemp products, such as plastics and paper.

Potential investors did not want to risk their capital on a government-controlled industry; farmers were weary of bearing the stigma of cultivating hemp due to its recent association in the public mind with marijuana. For these reasons, coupled with the powerful new restrictions, most hemp ventures began to decline in the fall of 1937.

On October 11, 1937, in the *Chicago Tribune,* reporter Frank Ridgeway (who had been directing a hemp project for the paper and had previously written a series of articles on the subject) spelled out the difficulties that the Marihuana Tax Act would create. The biggest problem was the transfer tax, which he felt was impractical and unrealistic. This is the section of the act that permitted the transfer of hemp (for example, from farmer to processor) only if the fiber-producing stalks were completely free of flowers and leaves.

The FBN enforced this new law as though all the foliage simply fell off the stalks during the retting process; however, farmers knew that some small quantity of leaves generally clung to the fibers, even after retting. The bureau's own research into hemp cultivation and processing should have made this fact clear to government officials, yet this regulatory agency maintained its restrictive stance, imposing upon both growers and processors the threat of criminal prosecution and confiscation of product.

In addition to the transfer, the Federal Bureau of Narcotics licensing procedure thwarted scores of legitimate hemp ventures. For example, in 1938, a group of hemp farmers near Lake Lillian, Minnesota, who had in 1934 and 1935 contracted to supply the Northwestern Hemp Corporation, hired Ohio attorney Olai A. Lende to represent their interests. These law-abiding farmers had applied for the hemp sales licenses stipulated by the Marihuana Tax Act, but had never received them. They had found a new buyer, Chempco, Inc., for their hemp, and were anxious to sell it. Lende contacted U.S. Senator Henkrik Shipsted of Minnesota regarding the FBN lack of response, and advocated the farmers' right to sell their legitimate crop. When Shipsted queried Commissioner Anslinger, the latter replied that the farmers were free to apply for a permit (which, of course, they already had), provided that their hemp was "substantially free of flowering tops and leaves."

Shortly thereafter, the prospective buyer, Chempco, Inc., closed down operations. There is no concrete evidence linking Chempco's closure to the government's interference. Yet, clearly, if the company was stonewalled from purchasing hemp fiber grown in previously established fields, it would be nearly impossible to find other farmers who would agree to supply the same crop. Finally, after several years of bureaucratic red tape, Attorney Lende stated in a letter:

> If I can find a market for the hemp, I have in mind to dispose of that hemp and tell Mr. Anslinger that he can go to the region below, and let him present the country with a spectacle of arresting half a thou-

Iowa farmers in the 1940s bundling hemp fibers utilizing specialized harvesting equipment.

sand farmers in Minnesota for selling an agricultural crop grown on their farms long before Congress ever thought of the Marihuana Act.

No one even knows if these farmers ever succeeded in finding a buyer for their hemp.

By the time the United States entered World War II, the resurgence of industrial hemp in America had faded, except for a few Wisconsin companies that traditionally had provided hemp fiber to the United States Navy for cordage and caulking. These enterprises were allowed to ignore the stipulations of the transfer tax, and stayed viable into the early 1950s. They were never targeted for potential violations, in contrast to the innovative firms that focused on chemurgic, cellulosic processes.

Government regulation and competition from the timber and petrochemical sectors had eliminated the profits to be derived by cultivating and processing hemp. Thus it attracted few supporters or market forces to encourage its continuation. Another more subtle factor in the failure of the American hemp industry was its rural, decentralized structure; its primary economic beneficiaries were farmers and small regional manufacturers. In that burgeoning era of big government and a few dominant companies, the centralized structure of the timber and petrochemical industries represented a more profitable approach. Trees and petroleum were in abundant supply, and were therefore cheaply priced. Furthermore, petroleum technologies promised a new age of synthetics, and Americans were attracted to the novelty of such products. Unfortunately, sixty years later the world now faces the consequences of deforestations, pollution, and financial consolidation that overreliance upon timber and petrochemical synthetics has brought.

"HEMP FOR VICTORY"

A fascinating "twist" in the U.S. government's hemp policy emerged in 1942, during World War II, when the importation of fibers for textiles and rope was curtailed. The Army and the USDA jointly released their "Hemp for Victory" campaign, featuring a film (now available on video) that rallied American farmers to grow hemp for wartime needs. Five years after passing the Marihuana Tax Act, the government was urging American farmers to grow hemp! The campaign certainly was not called "Marijuana for Victory," and it therefore signified that government officials still recognized the difference between the two varieties.

Scene aboard a Navy ship from the 1942 U.S. government film, Hemp for Victory.

German factory workers processing hemp fibers during World War II.

Legal research done in the 1990s to attempt to identify the USDA regulation that allowed the American hemp industry this brief resurgence has found no statute in place. Indeed, no additional regulation was needed in 1942, because the Marihuana Tax Act was designed to allow the legitimate hemp industry to continue. So, under the definition that is still on the books today, and with USDA encouragement, American farmers performed their patriotic duty by increasing hemp cultivation and processing, just as Harry Anslinger had assured Congress that they would be able to.

In 1942, the Kentucky Hemp Growers Cooperative Association was formed to supply the productive Kentucky hemp seed varieties for American growers (in its earlier formation, this organization only existed between 1942 and 1945). Farmers across the nation responded to the "Hemp for Victory" campaign, raising more than three hundred thousand acres of the crop during the war years. (For its part, Germany produced a book encouraging German farmers to plant hemp.)

When World War II ended, the U.S. government canceled virtually all hemp-farming permits, despite the many people who did not want this to happen. Guy M. Gillette of Iowa, a member of the Senate Committee on Agriculture and Forestry, was among those who voiced dissent:

> There are interests that do not want their markets usurped by the
> type of [hemp] industry that we have been trying to develop here,

and this committee has already given consideration to the presentation of legislation looking to the preservation and continuance of these industries that give promise to furnish an outlet for our surplus farm acreage and employment to our people in the postwar period.

The National Farmers Union called for "the widest use of hemp within the American market," declaring that farmers wanted "to determine the type of cooperative effort that could be organized to keep these mills in production and provide an outlet for a very satisfactory crop." Yet these and other protests against the renewed hemp ban were to no avail, and production declined drastically once again.

THE AMERICAN ERADICATION OF HEMP IN JAPAN

The end of World War II also brought a dramatic change to Japan's long, distinguished history of hemp cultivation. During the postwar Allied occupation, American and British troops were surprised to find hemp growing in abundance, as it had for centuries. In 1948, the Allies imposed a new national constitution that included the *Taima Torishimari Ho*, or Hemp Control Act.

Ironically, it was the Japanese Imperial Army's invasion of the Philippines in 1941 that had touched off the U.S. government's "Hemp for Victory" campaign to replace the Manila "hemp" (actually banana leafstalk fiber) used by the American military. Japan had relied on both domestic and Southeast Asian hemp crops during the war, to make uniforms, helmet linings, and other battle gear.

The commercial hemp plant was almost completely eradicated in Japan, as thousands of years of cultivation were buried under the avalanche of postwar change. Despite the efforts of the centralized government, however, some hemp was still cultivated until the mid-1950s (the plant also continued to grow wild). As in other countries where hemp became criminalized, eventually most Japanese farmers had no idea that this outlawed plant crop historically had supplied everything from birdseed to fine woven fabrics.

HEMP'S POSTWAR DECLINE AT HOME

In 1945, Congressman A. Willis Robertson of Virginia introduced H.R. 2348, addressing "Coverage of Certain Drugs Under the Federal Narcotic Laws." The bill basically granted the Treasury secretary the authority to determine whether "new" drugs should be considered opiates and subjected to the same controls as morphine and cocaine.

During hearings on this bill before the Senate Committee on Finance, Senator Robert M. La Follette of Wisconsin earned his place as one of the greatest hemp advocates since Thomas Jefferson. Responding to the concerns of some of his constituents (such as Matt Rens, a major hemp miller), La Follette questioned Deputy Commissioner William S. Wood of the FBN about the pressure Wood's agency was applying upon legitimate hemp farmers. For example, the FBN was withholding permits to transport hemp fibers for processing because the stalks contained some leaves. Referring to the Marihuana Tax Act as "a drastic piece of legislation," La Follette scolded Wood, and insisted that it was "perfectly clear" that the 1937 Senate had not been "putting the FBN in a position to wipe out this legitimate hemp industry." Wood assured the senator that his agency had no such intentions.

Unconvinced, Senator La Follette demanded that H.R. 2348 include specific language designed to protect the hemp industry; this was approved by the Finance Committee and became part of the bill. In presenting the amendment on the Senate floor, La Follette first summarized the history of hemp in America, calling it a "not large but very essential industry," and then concluded:

> I feel that it is important for the country to preserve the hemp industry. . . . I believe we have worked out an amendment that will not in any way weaken the power and control of the Bureau of Narcotics over traffic in marijuana, and at the same time I believe that the amendment, if it becomes law, will enable the legitimate producers of hemp to carry on their activities.

Several other senators echoed La Follette's concerns, and the amendment was approved. Matt Rens communicated his thanks in a letter to George Farrell of the USDA, dated July 27, 1945:

> Personally we are convinced that the hemp industry should be kept alive for our national welfare, and if this amendment had not been made and Mr. Anslinger had had his way the hemp industry would have been killed.

H.R. 2348 then moved back over to the House for consideration of the Senate amendment. The House initially rejected any changes, but the Senate insisted. Rather than allow his bill to die, Congressman

Robertson proposed a conference committee, whose conclusions he reported to his House colleagues:

> Mr. Speaker, the major purpose of this bill is to bring new synthetic drugs of a habit-forming character under our general narcotics acts and to treat them as we do opiates. . . . Both the House and the Senate were unanimous in support of the major provisions of the bill, . . . feeling that it is necessary to protect the hemp industry without imperiling our control of dangerous narcotics.

The House then voted to approve the Senate amendment, and another bill designed to control marijuana without threatening the legitimate hemp industry became law.

In the United States, the World War II hemp mills were shut down, and hemp soon retreated into oblivion. A few Wisconsin farmers continued to grow hemp to supply several firms such as the Matt Rens Hemp Company in Brandon, Wisconsin—the primary provider of rope for the United States Navy. But in 1957 this last American hemp processor, too, finally closed its doors. According to their catalog for that year, in 1958 the Columbian Rope Company of Auburn, New York, manufacturers of rope and twine, still made use of industrial hemp, as is shown in this quotation from the catalog: "Auburn Spring twine is made of just the right blend of fibers (principally American Hemp). This is a favorite twine of furniture manufacturers and the upholstery trade." But before long, the Columbian Rope Company was forced to rely upon imported hemp fibers.

ELSEWHERE, HEMP STILL FLOURISHES

While the post–World War II decline of American hemp was under way, worldwide hemp acreage surpassed a million acres in the late 1940s and early 1950s, as France, China, the Soviet Union, and the other communist countries of Eastern Europe in particular continued to grow and process hemp for numerous products. According to Dr. Janos Berenji of the Institute of Field and Vegetable Crops in Novi Sad, Yugoslavia (Serbia), the 1949 acreage in Yugoslavia alone totaled more than 238,000 acres. In many other countries not influenced by American legislation, hemp remained an important crop. In fact, France has grown hemp without interruption for at least six hundred years. But other countries were more affected by American hemp pol-

World Production of Hemp
by Principal Producing Countries
(Thousands of short tons)

Country	Period 1951	1950	1949	Average 1934–38
U.S.S.R.	a	a	a	223
Italy	72	73	78	99
Yugoslavia	33	33	77	51
Roumania	b	b	b	23
Korea	b	b	b	20
China—Manchuria	b	b	b	12
Hungary	b	b	b	12
Turkey	12	8	13	9
Chile	b	4	4	7
Poland	21	10	6	6
Spain	b	b	b	5
Japan	2	2	2	5
Czechoslovakia	b	b	b	5
France	5	5	5	4
Germany, total	b	b	b	4
Western	1	1	1	1
Eastern	b	b	b	3
Bulgaria	b	b	b	4
Syria	b	b	3	3
Sweden	b	1	1	c
United States	d	d	2	d
Other countries	b	b	b	3
TOTAL	390	360	387	495

[a]Not available, estimate included in total. Postwar program specified 1948 goal as nearly three-fourths prewar acreage ("A Survey of Soviet Russian Agriculture," Lazar Volin, U.S. Dept. of Agriculture, Washington, 1951); one estimate gives 112,000 short tons as production for 1949 and 140,000 short tons for 1950 ("Industrial Fibres," Intelligence Branch, Commonwealth Economic Commission. London: H.M. Stationery Office, 1952).

[b]Not available: estimate included in total.

[c]Included with "Other countries."

[d]Less than 500 tons.

SOURCES:

1951: "Monthly Bulletin of Agricultural Economics and Statistics," Food and Agriculture Organization of the United Nations, I: 1, Rome (1952).

1934–38, 1949, 1950: "Yearbook of Food and Agricultural Statistics," Vol. V, Part 1, Food and Agriculture Organization of the United Nations, Rome (1952).

icy. The United Kingdom outlawed hemp cultivation in 1971. In 1982, perhaps in emulation of the Reagan administration's "Just Say No" campaign, Germany passed similar legislation banning hemp. With the exception of a few acres in Chile, no country in Central or South America has grown hemp within the last several decades.

Yugoslavia has a long tradition of spinning and weaving hemp fibers.

AMERICAN HEMP POLICY INCONSISTENCIES SINCE 1960

In 1961, the U.S. Senate ratified and adopted the United Nations Single Convention on Narcotic Drugs, which attempted to coordinate the international effort to control increasing narcotics traffic, including marijuana. As directed in the Single Convention Treaty, the United States reorganized its controlled substance laws and transferred its enforcement mechanism from the Treasury Department to the Justice Department.

Importantly, however, one of the provisions of the U.N. Single Convention clearly states that "this Convention shall not apply to the cultivation of the Cannabis plant exclusively for industrial purposes (fiber and seed)." In fact, nearly all the hemp-producing countries in the world have signed the compact. Yet, although the United States also is a signatory, it has chosen to ignore—some would say violate—the agreement by prohibiting industrial hemp cultivation.

The Comprehensive Drug Abuse Prevention and Control Act

In 1970, Congress enacted our current federal drug-control law, the Comprehensive Drug Abuse Prevention and Control Act (CDAPCA). It replaced the Treasury Department's Federal Bureau of Narcotics with the Justice Department's new Drug Enforcement Administration (DEA) and declared that drugs "will be controlled in conformity with the treaty or other international agreement obligations" of the United States. The CDAPCA also repealed the Marihuana Tax Act of 1937 and replaced it with new control provisions.

The federal law cited by the DEA defines "marihuana" as:

> all parts of the plant *Cannabis sativa L.*, whether growing or not; the seeds thereof; the resin extracted from any part of such plant; and every compound, manufacture, salt, derivative, mixture, or preparation of such plant, its seeds or resin. Such term does not include the mature stalks of such plant, fiber produced from such stalks, oil or cake made from the seeds of such plant, any other compound, manufacture, salt, derivative, mixture, or preparation of such mature stalks (except the resin extracted therefrom), fiber, oil, or cake, or the sterilized seed of such plant which is incapable of germination.

Now, it might seem that the CDAPCA renders our previous discussion about the 1937 and 1945–46 Congresses irrelevant, since the laws crafted by those bodies were ultimately repealed. However, while the CDAPCA dramatically changed drug control strategy in the United States, it did not increase the number of substances controlled through previous legislation. In its report on the CDAPCA, Congress announced that the new law applied to those drugs "which by law or regulation have [already] been placed under control under existing law." The 1970 CDAPCA definition of "marihuana" is identical to the 1937 Marihuana Tax Act definition of "marihuana." But this time around, there was no discussion about changing or expanding that definition to address industrial hemp, which had been so painstakingly exempted from the earlier laws.

In fact, the House committee report on the CDAPCA explained that "marihuana would be prohibited except . . . for the emergency production of hemp." While the term "emergency" was not further discussed, this wording suggests that someone on the House committee remembered the "Hemp for Victory" campaign.

The CDAPCA delegates to the U.S. attorney general and the Health and Human Services (HHS) Department the power to add substances to the control list, provided that they are first evaluated for a variety of factors, including potential for abuse. The attorney general and HHS, in turn, delegates this authority to the DEA.

As is the case for any other federal agency, when the DEA adopts a substantive rule that would "create law" or impose "general, extra-statutory obligations," such as adding a new drug to the controlled substances list, it must first give notice to the public (in the Federal Register) and allow for comment. This public review process is mandated by federal law. But in spite of this requirement, the DEA has never given notice of its intent to include industrial hemp within the definition of marijuana. Likewise, no court has ever declared that federal marijuana laws have been extended to include industrial hemp.

Since the early 1990s, numerous legitimate farmers and processors have applied for permits to grow hemp. To date, the DEA has refused to grant permits to any industrial hemp applicants. There is no legal authority for the DEA's current position regarding industrial hemp, either from Congress or from any international treaty. Therefore, the DEA's position on hemp is unjustified. When DEA officials declare that industrial hemp and marijuana are the same thing and that both are illegal, as the agency has been doing since its inception, they are actually attempting to create law. This type of unreviewed "agency law" is the worst side effect of our increasingly bureaucratic system and is an anathema to true democracy. The individual states are free, and by right ought to be, to regulate industrial hemp within their borders as guaranteed by the Tenth Amendment of the Constitution of the United States: "The powers not delegated to the United States by the Constitution, nor prohibited by it to the States, are reserved to the States respectively, or to the people."

The illogic of the DEA's seemingly settled position—equating industrial hemp and marijuana—is betrayed by the internal inconsistency of the agency's actions. The DEA is required by the president to submit an annual report to Congress that identifies major drug-producing countries; thereafter, the United States subjects those nations to economic sanctions in order to discourage them. Despite the recent resurgence of the hemp industry in the United Kingdom, Australia, and Canada, along with continued hemp production in China, France, and Eastern Europe, the DEA has never reported any of these nations as marijuana-producing countries.

Elsewhere in the executive branch, the distinction between hemp and marijuana continues to be made. Hemp is recognized as an international commodity in both the North American Free Trade Agreement (NAFTA) and the General Agreement on Trades and Tariffs (GATT), and was listed as an "essential national resource" in President Clinton's June 3, 1994 Executive Order number 1291959 (listed in the Federal Register on page 29525).—the farthest designation from "opiate" imaginable.

Constitutionally, there is no question but that the regulatory framework surrounding industrial hemp cultivation in the individual American states ought to be handled by their respective state agricultural officials. Likewise, the USDA—not the DEA—should administer industrial hemp issues for the federal government. It is important to keep in mind that any American president has the authority to sign an executive order implementing the licensing of industrial hemp as recommended here. Industrial-hemp proponents do not want to change a single drug law; they only want the president to make a regulatory distinction between industrial hemp at 1 percent THC and marijuana.

THE THC ISSUE

In 1964, Dr. Raphael Mechoulam of Israel's Hebrew University identified THC as marijuana's psychoactive agent, thereby giving the world a means of scientifically differentiating between the *Cannabis* varieties.

Canada, Australia, and nations of the European Union (EU) now use gas chromatography to test the THC content of industrial hemp, and researchers are working to develop a device that will render accurate readings in the field. India, China, and Eastern European countries do not require any THC analysis of their hemp fields, though testing in those countries has shown that commercial varieties of industrial hemp contain less than 1 percent THC.

Marijuana, on the other hand, ranges from 3 percent to more than 15 percent THC. The establishment of a THC threshold to distinguish between industrial hemp and marijuana would create legal clarity based on a scientific process.

The EU gives hemp farmers a THC ceiling of 0.3 percent, down from an original 0.8 percent, as a criterion for receiving government subsidies. Some European hemp companies have suggested that France sought the lower 0.3 threshold to maintain its seed-supply domination in EU countries.

The case for a higher THC ceiling is supported by 1996 experiences in Germany: French seed stock planted there was found to exceed 0.3 percent quite frequently, either as a result of growing conditions or because second-generation seed was used. To develop high-yield varieties of fiber hemp, seed firms will need access to a wide range of genetic material. Recognizing that many varieties range from 0.1 percent to 1 percent THC, the North American Industrial Hemp Council (NAIHC) in 1996 recommended a 1 percent ceiling for seed cultivars. If forthcoming North American regulations dictate an unreasonably low 0.3 percent THC ceiling, the entire industry will be hampered.

Some have suggested that researchers should follow a different approach, selecting for changes in THC concentration to breed a THC-free hemp variety. A drawback of creating such cultivars may be that we could lose desirable genetic characteristics in the process.

In fact, hemp-seed breeders are calling for a regulatory framework that would allow them to work with seed varieties ranging up to a THC level of 2.5 percent. This would allow breeders to select for valuable genetic traits sometimes obtainable only from higher-THC hemp cultivars—ones still below the levels of marijuana. If such allowance isn't made, seed development may suffer because of the cumbersome red tape already associated with government permit processes throughout much of the world. Due to unnecessarily low current crop yields, over-restriction of seed breeders will further reduce hemp's ability to compete economically with cotton and timber.

LAW ENFORCEMENT CONCERNS

Many voices in the American law enforcement community have expressed concern that encouraging industrial hemp production in this country will increase the domestic supply of marijuana. While seemingly valid on the surface, these fears have not been borne out by the international experience. None of the twenty-nine hemp-producing nations have experienced any increased marijuana problems related to hemp cultivation. None of these hemp-producing nations are listed by the DEA as sources of illicit marijuana. Had all these fields such insidious potential, would they have gone unexploited all these years?

These same critics have claimed that unscrupulous persons could sneak into a hemp field and plant marijuana. Yet this notion also is a fallacy. Industrial hemp growers sow seeds in very dense bunches (300 to 500 plants per square meter), creating a thicket that is impossible to walk through. Imagine trying to make your way across your front lawn if every blade of grass were a quarter-inch thick and six to sixteen feet high! Dense planting causes the hemp to grow straight and tall, which produces the long, straight fibers best suited for processing. Marijuana growers, on the other hand, sow seeds at wide intervals (one to two plants per square meter), and force their plants to be bushy, with as many branches as possible, to increase the number of high-potency flowering ends. The two crops just don't work well together and are distinguishable even to the untrained eye.

More importantly, industrial hemp grown for fiber is harvested before it flowers, five to six weeks before marijuana growers would consider harvesting their crop. This means that any marijuana in a hemp field—already an unlikelihood—would be sent off to the fiber mill along with its cousins, and would not yet have sprouted any intoxicating flowers or buds. Marijuana growers are much likelier to continue to prefer the long, open rows of a cornfield to provide cover for their illicit crop. In fact, pollen from industrial hemp being grown for seed will suppress the THC production of marijuana crops within an approximate ten-mile radius. This is why clandestine marijuana growers are so vociferously opposed to the relegalization of industrial hemp, which they know will render their hidden crops much less profitable.

It is puzzling that the DEA still expresses concern over a perceived problem that their colleagues in twenty-nine other countries have reported no trouble handling. Initially, European law enforcement officials also had strong misgivings about crop theft and camouflaging of marijuana. Hemp permit requirements in the EU include

- background checks on growers;
- mapping of hemp plots;
- locating plots in places where they will be "invisible,"

with limited road access and screening crops
(for example, corn);
- inspections of fields.

In 1996, plants from more than three hundred German fields (over 50 percent of the hemp crop) were sampled; no illegal drug cultivation was detected. European officials in agriculture and health ministries now consider the risk of drug abuse of low-THC hemp to be negligible; there has been very limited crop theft. The Europeans have seen that hemp farmers are not prone to grow marijuana and risk their livelihoods. Camouflaging individual drug plants is a conceivable risk but is not expected to increase the supply of drugs. According to the German nova Institute, a scientific consulting organization, government officials in Europe are satisfied with the controls presently in place.

Are we to assume, as opponents of industrial hemp appear to do, that the American law enforcement community is less competent than its foreign counterparts? I personally believe our police to be quite capable of carrying out a carefully conceived and implemented law enforcement program that addresses the need to control illegal drugs while permitting the cultivation of industrial hemp.

Since hemp is an agricultural crop, it is now appropriate to shift the main regulatory responsibility for hemp from the Drug Enforcement Agency to the Department of Agriculture. This is especially important in light of the fact that the DEA's predecessor agency, the Federal Bureau of Narcotics, was responsible for effectively hamstringing the hemp industry. Americans interested in supporting the hemp industry should contact the White House along with their federal representatives in the House and Senate and request help in bringing about this transfer. With a few pen strokes, President Clinton has the authority to enact an executive order empowering the Department of Agriculture to develop a regulatory framework for the licensing of hemp farmers. Industrial-hemp proponents do not want to change any laws; they only want the president to make a regulatory distinction between industrial hemp, at 1 percent THC, and marijuana.

It is ironic that a fiber so useful to humanity is also associated in the public mind with a forbidden drug that has a huge black-market demand. Proponents of industrial hemp have made a clear distinction between "the rope and the dope," and they do not advocate drug-law changes. Opponents of industrial hemp, including the law-enforce-

ment community, erroneously tell legislators and the general public
that hemp and marijuana are one and the same. Advocates of medical
marijuana and drug legalization often associate hemp with marijuana,
thus further clouding the issue. In addition, the emotionalism of the
"war on drugs" and the huge profits generated by financial institutions from the laundering of drug money (read the eye-opening book
*The Laundrymen: Inside Money Laundering, The World's Third-
Largest Business* by Jeffrey Robinson) make for a frequently irra-
tional, highly charged debate on the subject of hemp.

HEMP'S GLOBAL STATUS ENTERING THE 1990S

By the early 1990s, worldwide hemp acreage had fallen to its lowest
levels in recorded history. Virtually every Western country except
France had banned or fatally restricted hemp cultivation, and Eastern
Europe had lost its major hemp consumer due to the collapse of the
Soviet Union in the late 1980s. Outside of France and China, hemp had
all but disappeared as a fiber crop by the early 1990s.

The international "Hemp Wall" began to crumble in 1993, when
the United Kingdom lifted its ban on domestic hemp cultivation;
Canada and Australia permitted test plantings in 1994. A handful of
activist-entrepreneurs in Canada, Austria, Germany, Australia, the
Netherlands, the United States, and the United Kingdom began to import hemp products from China and Eastern Europe. The Chinese,
Hungarians, and Romanians were at first perplexed about why these
Westerners were so interested in hemp. The mill and factory owners
of these developing countries viewed hemp as an old-fashioned relic
of their own ancient cultures—perhaps even an anachronism that
should be abandoned in favor of modern synthetics. But since the international hemp traders were willing to write checks, several old factories began producing rather poor-quality textile fabrics. Other entrepreneurial companies imported hemp seed and oil for food and for
cosmetic applications. In the early 1990s, dozens of small, pioneering
companies in the United States such as Ecolution, Terra Pak, Hemp Essentials, Artisan Weavers, American Hemp Mercantile, Hemp Textiles
International, the Coalition for Hemp, the Ohio Hempery, Sharon's
Finest (now The Rella Good Cheese Company, Two Star Dog, and
Hempstead created new businesses by focusing on specialized products and markets.

In 1994, the Hempstead Company received a USDA license to plant industrial hemp at the agency's field station in California's Imperial Valley. Unfortunately, just before harvest, state officials plowed the crop under upon orders from the office of Attorney General Dan Lundgren in Sacramento.

Recent American Legislative and Political Efforts

Nevertheless, American farmers have continued to agitate for the right to resume the large-scale cultivation of industrial hemp. In October 1994, the Kentucky Hemp Growers Cooperative Association (KHGCA), dormant since 1945, was reactivated by some seventy Kentucky farmers and civic leaders. Following Bioresource Hemp, the March 1995 landmark international hemp conference in Germany, the Wisconsin Department of Agriculture sponsored several key meetings of farmers, policymakers, businesspeople, environmentalists, and researchers. These meetings in turn spawned the North American Industrial Hemp Council (NAIHC), a broad-based educational and advocacy coalition, in early 1996 (see chapter 4 and the Hemp Resources list for further information about international hemp conferences and organizations).

In January 1996, KHGCA president Andrew Graves introduced a resolution to the fifty-one (including Puerto Rico) state delegates of the American Farm Bureau, whose members number more than four and one-half million. Graves's resolution supported planting research plots of industrial hemp, and all of the state chapters voted for it, generating considerable bipartisan support for renewed exploration of fiber hemp. Domestic cultivation of industrial hemp also has attracted support from the American public. A 1995 University of Kentucky poll found that 77 percent of Kentucky citizens support the licensing of hemp farmers; Vermont residents expressed comparable approval in a 1996 statewide poll (conducted by the University of Vermont). And in this period of reduced federal funding, Native American tribes such as the Navajo Nation are very interested in developing hemp-farming ventures in order to create a more self-sustaining tribal economy.

With momentum growing to revise industrial hemp policy, the latter 1990s have seen mounting legislative endeavors within the United States. In 1995, one state—Colorado—introduced pro-hemp legislation. In 1996, four states did so: Colorado, Hawaii, Vermont, and Missouri. In 1997, eight states introduced legislation: Colorado,

The Kentucky Hemp Growers Cooperative Association

The Kentucky Hemp Growers Cooperative Association (KHGCA), originally formed in 1942 in response to World War II's "Hemp for Victory" effort, was reactivated in October 1994. The well-respected Jake Graves, Sr., a retired banker and fifth-generation Kentucky hemp farmer, serves as chairman of the KHGCA, which now consists of more than seventy members interested in growing hemp in Kentucky. In an ironic twist, many tobacco farmers want to grow hemp—a crop that isn't a drug—in their efforts to find an alternative to tobacco—a harmful and addictive drug.

Now as before, the KHGCA serves to establish and promote market equity among industrial hemp suppliers, processors, and manufacturers. However, this time around, the crop needs considerably more educational efforts on its behalf, so the KHGCA is at work here, too.

Joe W. Hickey, Sr., a hemp farmer's grandson and the association's executive director, has phoned, faxed, and e-mailed information about industrial hemp to tens of thousands of contacts around the world. He says, "The hardest challenge has been educating the public and demonstrating that industrial hemp is not, and never was, marijuana." He also has had to dispel the myth "that industrial hemp is the great savior of the American farmer."

Like other hemp advocates, Hickey sees his efforts as an opportunity to help farmers diversify and develop their operations in environmentally sound fashion, allowing them to receive fair returns for their efforts, become more independent, and remain on the land in an era of otherwise disheartening rural economic trends.

Missouri, North Dakota, Iowa, Minnesota, Virginia, Kansas, and Hawaii; two more—Oregon and Wisconsin—wrote bills that were not introduced.

Of all this legislation, three bills were passed permitting research (not planting), in the states of Hawaii, Vermont, and North Dakota. Working against each of the bills was considerable law-enforcement opposition that often intimidated legislators. Wisconsin police officers telephoned legislators, their spouses, and their pastors to say that the legislators were promoting marijuana use. The sheriff and the district attorney of Marathon County, where the officers are based, later sent a letter to the Industrial Hemp Forum, addressing it to the "Industrial Marijuana Forum." This letter inaccurately described the hemp bill as legalizing the growing of marijuana.

In Minnesota, a Ms. Jeanette McDougal, representing Drug Watch International, testified inaccurately to Minnesota legislators that the American Farm Bureau's Resolution 65 (supporting industrial hemp) had been withdrawn and rejected. Her testimony was, if not an outright lie, a deliberate misstatement of fact.

While many states were working to develop new job and business opportunities based on the processing of hemp fibers, Oklahoma took a different tack: The governor signed legislation encouraging aerial spraying of herbicides on feral hemp stands that have survived from the World War II "Hemp for Victory" plantings. Nevertheless, many lawmakers had the courage to stand up in favor of hemp. "Thomas Jefferson and George Washington were both industrial-hemp producers," declared North Dakota Republican representative David Monson. And Ned Breathitt, Kentucky's former governor and current chair of the University of Kentucky Board of Trustees, came out strongly in favor of legalizing industrial hemp.

The following is a more in-depth look at political activities regarding hemp in five states.

Vermont

In 1996, State Representative Fred Maslack introduced a bill that outlined the legislature's intent to grant hemp cultivation licenses to universities and private farmers. Said Maslack (a Republican himself):

> This is a real Republican issue. All we are proposing is to get government out of the way of a creative, productive endeavor. Looking at the big picture, this is an indigenous, sustainable, agricultural economic development.

His bill was supported by the State Agricultural Committee, the Vermont Farm Bureau, the Association of Vermont Dairy Farmers, and nearly 80 percent of the Vermont residents who responded to a statewide poll. After Governor Howard Dean threatened a veto, a compromise bill passed, directing the University of Vermont to conduct a study on hemp (see chapter 5). Yet the key passage that was concerned with new allowances for commercial cultivation of hemp had been deleted to insure the bill's passage.

Hawaii

Thousands of acres of fertile land lie fallow here because of the decline of the local sugar industry. Astute politicians such as Republican representative Cynthia Theilen know that Hawaii sorely needs a strong agricultural commodity with which to stabilize its tourist-based economy. In 1996, both houses of the Hawaii legislature passed resolutions H.R. 71 and H.C.R. 63 to study the cultivation of industrial hemp. The Summer 1996 issue of the magazine *HempWorld* quoted Representative David Tarnas:

Industrial hemp can supply a raw agricultural commodity that will support former sugar plantation workers as well as numerous cottage industries for textiles, clothing, and food products. The potential for fuel production is also very attractive, since we import all our fuel. Industrial hemp should be allowed to prove itself as a successful commodity here in Hawaii as it is doing in other places around the world.

Colorado

In 1995, Senator Lloyd Casey introduced the Industrial Hemp Production Act, which died in committee. The 1996 bill then introduced by Casey would have provided for the regulated cultivation of up to forty acres of industrial hemp by Colorado farmers for agricultural, commercial, and scientific purposes. Proponents of the bill worked closely with the Rocky Mountain Division of the DEA, educating the agency about the differences between hemp and marijuana.

Just two days before the state house hearings on the bill, the Denver Police entered Casey's capitol office and seized a legal bale of Canadian-grown hemp. No charges were filed, but the bale was not returned. Amid frantic lobbying by nonfarm interests, antidrug groups, and law enforcement, all of which used misinformation and scare tactics, the bill, which actually had widespread support, passed the state senate but was killed once more by the House Agriculture Committee. Nevertheless, Colorado's bold attempt created national attention and set the stage for several other states to act. Former state senator Lloyd Casey has now founded the Agricultural Hemp Association to lobby and educate across the nation.

Missouri

Stacey and Boyd Vancil of the Oxford Hemp Exchange in Poplar Bluff, Missouri, have coordinated a strong pro-hemp educational campaign throughout the state. In 1996, the Missouri senate passed a 22–6 roll call vote supporting hemp research. In 1997, Missouri came closest of any state to passing a bill that would allow farmers to grow industrial hemp. The hemp legislation passed strongly, 2–1 in the Senate Agriculture Committee and 2–1 in the House Agriculture Committee, before being derailed in the full house.

Adidas Launches "The Hemp"

In 1996, the Adidas shoe company introduced a model whose upper was made of the hemp. As this was one of industrial hemp's first major American consumer-marketing efforts, it drew severe criticism from Director Lee Brown of the White House's Office of National Drug Control Policy. Brown asked Adidas to withdraw the shoe from the marketplace, a request that was denied. "I don't believe you will encounter anyone smoking our shoes any time soon," responded Adidas president Steve Wynne.

The Adidas–White House hemp controversy drew considerable media coverage, such as the following excerpts from an editorial entitled "Things Went Better With . . ." that appeared in California's *Ventura Starfree Press* (February 11, 1996):

> Anti-drug organizations have cried foul to no effect, in an attempt to prevent the Adidas shoe company from releasing its newest product, "The Hemp," a shoe named for the material from which much of it is fashioned.
>
> While it is likely the marketing staff at Adidas believes the name "Hemp" will enhance the casual shoe's appeal in the youth market, it is also difficult to argue that it is wrong for a manufacturer to identify a product by the material from which it is made. Consumers certainly use such descriptions generically: cashmere sweaters, silk sheets, canvas sneakers, polyester suit, patent leather shoes.
>
> Once upon a time, a Georgia-based company made a product that included the use of coca plants, whose dried leaves are also used in the production of cocaine. It called its product "Coke." That trademark worked out all right, and no one suggested that it promotes the use of illegal drugs.

Kentucky

In late May 1996, actor and environmentalist Woody Harrelson was in Kentucky for the KHGCA conference in Lexington. He has been outspoken about government policies that allow for the clear-cutting of old-growth forests, and sees hemp as practical way to lessen the need for timber. He invited the national media and the local sheriff's office to watch as he planted four hemp seeds (five would have constituted a felony) on one square foot of land (lest the property be seized under drug forfeiture laws) that he had bought in Boonesville, Kentucky. Thanks to his celebrity status, his story made headlines as he was arrested.

Harrelson's case, *Commonwealth v. Woody Harrelson*, received extensive press coverage. Lee District Judge Ralph McClanahan stated that the definition of marijuana, under which Harrelson was charged, "is constitutionally defective due to its over-broad application. . . ."

"There's a judge with vision," Harrelson said after the court ruling. "I couldn't be happier. This is a great day." In July 1997, Lee Circuit Judge William Trude upheld the earlier court ruling. As this book goes to press, the prosecution is appealing the ruling yet again, this time in the Kentucky State Appellate Court.

In summer of 1997, the Kentucky legislature scheduled hearings to discuss the feasibility of industrial hemp.

Woody Harrelson.

M. Patterson-Thomas, Lexington Herald-Leader

Hemp Enters the Limelight

An unrecognized benefit to local communities is that citizens are now seeing firsthand how their elected and unelected officials vote or act on such issues. A "Yes" vote for hemp helps farmers and industry, creates jobs, and protects the environment; a "No" vote forgoes these opportunities and permits law enforcement to continue to create policy on the fly. Perhaps after a few more years of these intense debates in their legislatures, local citizens will vote out politicians who are not representing their community values.

In any event, there will be ongoing and increasing legislative efforts to restore industrial hemp to its once prominent role in agricultural economics. Robert Winter of Colorado's Weld County Farm Bureau points out, "It will be done by someone sometime in the near future. The question is, which state will take the leadership role?" The state that is the first to encourage commercial hemp activity should expect an economic boost as businesses move in to take advantage of a supportive local environment.

As U.S. farmers and rural communities face severe economic problems, it becomes more and more difficult to justify the current ban on domestic hemp cultivation. While twenty-nine other countries raise hemp and snap up patents, profits, and market share, how long will the American government persist in maintaining unnecessary agricultural regulations? In this era of the global economy, can the United States afford to stand alone among developed nations in impeding the progress of industrial hemp?

Business Profile

Two Star Dog

Two Star Dog is a hemp clothing company that has recently expanded into a line of hemp-oil bodycare products. Steven Boutros, owner and president of Two Star Dog, explains the original motivation for founding this company:

> Our organization was started with the philosophy of blending modern manufacturing techniques with a concentration on renewable resources and processes that have little or no negative environmental impact. When we learned about hemp, its versatility, and its ability to actually improve the soil it's grown on, as well as its potential to replace environmentally damaging industries such as petroleum and timber, we saw that this is an industry with a real future. Given the consumer demand for natural and environmentally safe products and manufacturing processes, this was a niche that we felt comfortable working in, and one that was wide open for exploration and opportunity.

Like other hemp entrepreneurs, Boutrous must cope with various roadblocks, but he is optimistic about the positive outcomes of overcoming these obstacles:

At present, the challenge of finding competitively priced raw materials will probably remain a constant to deal with. Eventually, domestic production could not only bring down the price by reducing transportation costs, but the inflated price of hemp raw materials could fall to an acceptable level on a world-wide basis. Added competition, combined with more refined processes, could increase hemp's viability as a competitive raw material. Employing our farmers and factory workers could close the loop and invigorate our economy with a new infrastructure based on hemp and other renewable resources. This would establish a new economic matrix for industrial growth without environmental destruction.

Two Star Dog
1370 10th Street
Berkeley, CA 94710
Telephone: (510) 525-1100
Fax: (510) 525-8602

Two Star Dog designer and partner Stella Carakasi looks sharp in her elegant hemp dress.

In Ontario in August, 1994, just 70 days after planting, Joe Stobel of Hempline admires a 10-foot hemp plant, grown for the first time in over forty years under a research license issued by the Canadian government.

The United States is an island of denial
in a sea of acceptance.

ERWIN SHOLTS
WISCONSIN DEPARTMENT OF AGRICULTURE

Around the Globe with Hemp Today

CHAPTER FOUR

THE CRACKS in the Hemp Wall are widening rapidly as twenty-nine countries, including most of the world's major industrialized nations, expand their hemp industries to gain economic and environmental benefits. This chapter provides an overview of the current status of hemp production around the globe, followed by a recap of recent important symposia. In addition to the countries covered here, Nepal, Ireland, Portugal, Jamaica, and Thailand produce hemp, as do most nations of the former Soviet bloc.

Australia

In 1994, government officials permitted one farmer to grow hemp on a small-scale research plot. Today the crop is being grown for research purposes in five of the seven Australian states.

Several entrepreneurial firms, including Australian Hemp Products, are now manufacturing and marketing products using imported hemp. Imported hemp baling twine is in demand by the wool industry, which prefers it to polypropylene cordage as a strong, noncontaminating natural fiber that will take dyes. Producers of particleboard are showing great interest in hemp's potential as a substitute for wood chips, and in 1997 the country's first domestic hemp-seed oil will be produced. An informative, one-hour Australian TV documentary, *The Billion-Dollar Crop*, is available on video. The combination of media interest and strong support from both farmers and the public is expected to lead to full-scale agricultural production by 1998.

Austria

As in France, the farming of hemp (*hanf*) has never been prohibited in Austria. Yet from the late 1950s until 1995, there was little interest among Austrian farmers in growing the crop. In 1995, with assistance from the Austrian Hemp Institute, some farmers planted trial hemp fields. These farmers are growing crops for both fiber and seed. Numerous institutions within the country, including the Department for Organic Agriculture at Vienna's University of Agriculture, Forestry, and Renewable Resources, are now conducting research into hemp's potential role in organic farming systems. It is likely that the world's first certified organic hemp products will be developed by the ecology-minded Austrians.

Canada

Canada was a major hemp-growing region until the twentieth century, when that country followed the example of the United States and prohibited hemp production. In 1994, Canada issued its first license in more than forty years to the persistent Joe Strobel and Geof Kime, founders of Hempline, Inc. They were allowed to plant ten acres of industrial hemp in Ontario, on land that had previously been cultivated for tobacco.

In December 1994, Agriculture and Agri-Food Canada, the federal agriculture department, devoted its *Bi-weekly Bulletin* (Vol.7, No. 23) to hemp farming, and printed it on hemp paper. By 1995, a total of thirty-five acres of hemp trials had been planted in the provinces of Ontario, Manitoba, Saskatchewan, and Alberta. Although the old law still stood on the books, making it illegal to possess any form of *Cannabis sativa*, dozens of hemp manufacturers and retail outlets began to appear in major cities across Canada. They relied on imported hemp to circumvent the ban still in effect.

Geof G. Kime

Hempline used a simple tractor and mower for harvesting hemp at their 10-acre test plot in Ontario, Canada.

By early 1997, Bill C–8 had become law to clarify the legality of hemp-fiber products in Canada. While the new legislation did not spell out any licensing procedures, it reclassified hemp stalks as a legal, tradable agricultural commodity rather than a narcotic, and gave the Ministry of Health the option to develop a regulatory framework. Additionally, to make the bill more palatable to its opponents, much harsher penalties for marijuana use and cultivation were included.

Although the wheels of government turn slowly, many Canadians have been assured that the Ministry of Health will have licensing regulations in place for the 1998 growing season.

In 1997, Hempline is processing one hundred acres of hemp, utilizing its newly developed fiber-separation technology. Most of this crop is being sold to manufacturers in the United States. Future Canadian exports of hemp products to the United States are expected to increase rapidly.

Chile

The Spanish monarchy of the 1400s considered hemp (*cañamo*) to be such a vital crop that it required New World colonial growers to supply it. Hemp cultivation in Chile therefore has a five-hundred-year tradition. Spanish hemp cultivars have flourished in Chile, due to the two countries' similar latitudes. Nevertheless, while it remains legal to grow hemp in Chile today, this is only a very small niche crop. There is, however, renewed interest in expanding the Chilean hemp industry for export purposes.

China

As discussed in chapter 2, China has been growing hemp (*ma*) for at least six thousand years, and is by far the world's largest consumer and exporter of hemp seed, paper, and textiles. The country's total annual hemp production exceeds one hundred thousand acres. The majority of pure hemp and hemp-blend textiles in Western markets originate in China, and even larger amounts of these fabrics stay there for domestic use or are sold to other Asian markets. In 1996, a mill in Dong Ping, Shandong—the biggest hemp mill in the world—invested about $11 million in hemp-textile spinning equipment. The Dutch company Naturetex is the exclusive distributor for this mill, which supplies fabric to hundreds of companies internationally, including Giorgio Armani.

China also has more than a dozen particleboard factories that use hemp fibers to manufacture construction materials for domestic use.

With its vast natural resources, labor pool, and consumer markets, China will continue to have a major influence on the future of the global hemp industry.

Denmark

Following several decades of inactivity with regard to hemp (*hamp*) cultivation, Denmark planted its first modern hemp trials in 1997. The Danish Hemp Society is working to reestablish hemp in this country renowned for its ecological leadership. In 1997, a major government-funded study highlighted the economic benefits of converting Danish agricultural practices to organic methods.

Finland

The cultivation of hemp (*hampu*) is a long-standing tradition in Finland, where hemp seeds dating back to A.D. 1000 have been found buried on the island of Ahvenanmaa. But since the 1960s, hemp has fallen out of favor in Finnish agriculture, as in most other countries. In the summer of 1995, as hemp was regaining global attention, the Hankasalmi Hemp Project planted several small test plots; from some of the retting fiber thus produced, children prepared handmade paper as part of a school project.

France

France has grown hemp (*chanvre*) continuously for at least six centuries, and in 1996 harvested more than ten thousand tons of industrial hemp.

Kimberly-Clark Corporation's French operation manufactures specialty hemp papers, for such applications as Bibles and cigarettes. Several French companies today are combining hemp fibers and lime to make a lightweight concrete/plaster with Earth-friendly insulating and pest-resistant properties. It can be used in both new home construction and remodeling.

La Chanvrière de l'Aube

A new French manor with hemp plaster on its outside walls.

Germany

The cultivation and use of hemp (*hanf*) in Germany had been in steady decline even prior to the ban placed on hemp farming in 1982. The 1993 publication of the book *Hanf* by Jack Herer, Mathias Brockers, and Michael Karus sparked renewed interest among the media and general public in "bioresource hemp." Initially, Germany's small yet innovative and fast-growing domestic hemp industry focused on the design, manufacture, and distribution of long-fiber apparel and furniture textiles from imported fabric.

The product palette has since grown rapidly, and now includes a wide range of paints, detergents, tasty foods, and bodycare products made from hemp seed and oil, as well as a range of paper products. In 1996, total German retail sales of hemp products surpassed $20 million.

In March 1995, the nova Institute held Bioresource Hemp, the first scientific symposium on hemp, which sparked international communication among researchers and groups already involved in hemp. At the same time, the federal government, driven by extensive media coverage, public opinion in favor of hemp, and pressure from farmers, began to move toward relegalizing the planting of low-THC hemp. This relegalization process required only the adoption of applicable European Union guidelines into national law. It was further eased by the fact that several other EU countries were already growing hemp, and by a hefty EU subsidy of $400 per acre.

Following Germany's relegalization of industrial hemp in March 1996, German farmers grew and harvested hemp on 3,500 acres. Fifty percent of this acreage was farmed under contract with emerging hemp-processing firms, including 750 acres contracted by HempFlax and 280 by Badische Faseraufbereitung (BaFa), companies which guaranteed purchase of the crop, bought the increasingly expensive French seeds, and gave growing and harvesting support to the farmers.

The German growers' initial experience with hemp cultivation has gone well. Crops grown on good soils achieved the expected weed suppression and good yields of 3.5–4.4 tons per acre. As anticipated, the shortage of adapted harvesting equipment did create some challenges for the farmers, who tried and modified several different approaches, including bar mowers, mower/conditioners, and the Hemp-Flax harvester.

In 1996, the nova Institute conducted its Hemp Product Line Project to evaluate market opportunities for domestic hemp. Results indicated that a number of hemp-based product lines (cottonized hemp textiles, specialty paper, composites, carpeting, and animal bedding) may compete technically and economically with conventional products, and would justify hemp farming on fifty thousand or more acres by early in the next decade.

Hungary

Until the collapse of the Soviet bloc, Hungary was a major supplier of finished hemp (*kender*) products to the Soviet Union. The Hungarians produce hemp twine, rope, carpet, fabric, and paper for export; much of this goes to the United States.

Hungarian hemp cultivars, which produce the world's highest yields of fiber and seed, also are exported worldwide. Dr. Ivan Bócsa, coauthor of *The Cultivation of Hemp* and the world's leading hemp breeder, conducts research at the GATE Agricultural Research Institute, where he has developed hemp-seed varieties of outstanding quality.

India

As in other parts of Asia, large stands of naturalized *Cannabis* are found throughout many regions of India. The fiber is used locally for the production of cordage and crude textiles, and the seeds are sometimes pressed for oil. Although very little hemp is intentionally grown as a fiber crop in India, peasant craft items and clothing fashioned from Indian hemp do find their way to Western markets. Pilot programs are currently studying the potential use of large tracts of naturalized *Cannabis* for the production of textiles, paper pulp, and building materials.

Japan

Hemp (*taima*) was cultivated throughout a long, distinguished history in Japan. This tradition began to disappear with the passage of the 1948 Allied-imposed national constitution, which included the *Taima Torishimari Ho*, or Hemp Control Act. After Emperor Hirohito died in 1989, his son's coronation as the next living embodiment of God included a special Shinto ritual. Religious tradition required that the new emperor wear hemp garments. However, these had become unavailable over the course of Hirohito's long reign, due to the hemp ban. A group of Shinto farmers on Shikoku saved the day: They had planned ahead for the coronation and raised an illegal crop in secret, and now presented the new emperor with clothes made of pure Shikoku hemp. On Shikoku (the smallest of Japan's four main islands), a bit of hemp is now cultivated for the exclusive use of the imperial family.

In the early 1970s, the first modern hemp symposium was held at Kyoto University and a court challenge was filed to argue that the ban was unconstitutional. The hemp movement became a broader struggle to overturn not only the ban but also the pressing thumb of American influence and military presence.

In Japan—as in the United States—current law actually allows farmers to apply for permits to cultivate hemp. Yet making such an application turns out to be a lengthy, futile process, since the government simply does not issue permits.

Meanwhile, Japanese artists continue to use imported hemp to make traditional sheer woven cloth, hand-dyed curtains and screens, paintings, and quilts. Ground hemp seed remains in the Japanese diet in the *schichimi* (seven spices) used for flavoring noodles.

It has recently come to public attention that the Hemp Control Act was instituted for a period of only fifty years, so the law should be up for review in 1998. Perhaps this will bring about the reemergence of a crop as traditional there as rice, silk, or bamboo.

The Netherlands

Recently, the Dutch government funded an extensive four-year study to evaluate and test the practical aspects of growing hemp (*hennep*)

Square hemp bales being loaded at the HempFlax factory.

and processing it into pulp for paper production. A respected Dutch researcher, Hayo van der Werf of the International Hemp Association, has written a treatise on hemp cultivation entitled *The Crop Physiology of Fibre Hemp*.

Hemp farming for a variety of purposes is on the rise in the Netherlands. In 1997, the Dutch firm HempFlax planted approximately four thousand acres of hemp for use as animal bedding, composite board, and pulp with which to strengthen recycled paper.

HempFlax also has developed innovative harvesting and processing equipment, and is working with European manufacturers to create new markets in such product areas as paper, textiles, and molded products.

Poland

Like Hungary and Romania, Poland has a long hemp-growing history. Poland currently grows hemp (*penek*) for textiles and manufactures hemp particleboard for building materials. The country's Institute of Natural Fibers, a leading agricultural fibers research organization, has demonstrated the use of hemp and flax cultivation as a means to cleanse soils contaminated by heavy metals (see chapter 6, and the 1995 report *Bioresource Hemp: Proceedings of the Symposium*).

Romania

Though rural Romanians still make use of the crop in traditional ways, Romania currently is Europe's largest commercial producer of hemp textiles, and the world's second largest exporter (after China). Over the past several years, Romania has based much of its textile production on surpluses of stalks left over after the breakup of the Soviet bloc. Since the country's own processing facilities are rather antiquated, new investments by companies such as Austrian-based Rohemp are helping Romania to expand its export opportunities.

Russia

Until the 1900s, Russia was the world's largest cultivator, manufacturer, and exporter of hemp (*konopli*), employing much of its immense peasant labor force in that industry. The Vavilov Research Institute in

Saint Petersburg maintains the most extensive hemp-germ-plasm collection in the world, one that includes many varieties not found in other gene banks.

South Africa

Industrial hemp has been unknown to South African farmers, but in late 1995 the State Department of Health granted permission to the South African Hemp Company to conduct trials at the Tobacco and Cotton Research Institute in Rustenburg. The first plantings failed for several reasons, including a prolonged drought and use of the wrong varieties for that latitude. Further research is required before hemp can be adapted to South Africa's predominately arid landscape.

Spain

The cultivation of hemp in Spain has continued uninterrupted for centuries. Spain currently is the sole exporter of hemp (cañamo) pulp for specialty papers. Several decades ago, the country's other hemp exports included rope and textiles. Now that global demand for hemp products is again on the rise, we can expect Spain's production to expand.

The United Kingdom

In 1992, French-grown hemp fiber was being sold extensively in the United Kingdom as animal bedding. British farmers—prevented since 1971 from raising hemp—began to complain because French farmers were under no such prohibition, and so had gained an unfair market advantage. The following year, an agricultural trading firm in Essex, England, formed a division called Hemcore. Employing the dual strategy of threatening a legal suit and quietly lobbying the government, this company was successful in getting the British Home Office to lift the domestic ban and issue permits for hemp cultivation.

Working with local farmers, Hemcore planted the first new English crop in 1993; subsequent crops have also been grown in Ireland and Scotland. Since then, hemp acreage in England alone has increased by roughly 50 percent every year (mainly under contract with Hemcore), reaching six thousand acres in 1997.

The government has been supportive, giving grants of £100,000 (US $161,000) to Hemcore and the Natural Fibers Organisation to develop new markets for hemp and flax fibers. In 1995, Bioregional Development Group, in partnership with farmers and industry, succeeded in producing the first U.K.-grown, machine-processed, 100-percent-hemp fabrics of this century. Numerous other organizations are working toward the reestablishment of the processing and production of hemp textiles, paper, and other products. One exciting endeavor is that of Lister and Company, the U.K.'s largest knitter of woolen rugs, which has introduced a line of hemp and hemp–velvet textiles for such designers as Ralph Lauren and Tommy Hilfiger.

The United States

The federal government has not granted any permits for large-scale farming of "true hemp" in more than forty years. However, increasing interest in the use of environmentally friendly products has created a burgeoning demand for hemp goods. In the early 1990s, a few American activist–entrepreneurs began to import hemp products from China and Eastern Europe. Recently, major designers including Ralph Lauren, Giorgio Armani, and Calvin Klein are selling hemp fashions in high-end department stores. Total gross retail hemp-product sales for all U.S. business were estimated to be $50 million in 1996.

Given renewed consumer interest and the fact that supplies are still limited (all raw and processed hemp material must be imported), American business costs are artificially high. As a groundswell of commercial, agricultural, and other interests continues to pressure the government for the right to cultivate hemp, such major trading partners as China, France, Germany, and the United Kingdom export tens of millions of dollars worth of raw hemp and related goods to the United States. In coming years, free-market traders very likely will keep jumping over the Hemp Wall to deliver these products. American consumers are voting with their dollars, and the race is on to sell to this expanding market.

Yugoslavia

Hemp (in Serbo-Croation, *konoplja*) has been an important crop in this area of Eastern Europe since at least the fifteenth century. The most important hemp production area in the Yugoslavian region traditionally has been the northeast part that neighbors with Hungary and Romania—the province of Vojvodina. In the town of Odzaci, a unique stock market of hemp straw was operated until the 1950s, which influenced hemp prices in countries as far away as Italy and England. A hemp research station established in 1952 in Backi Petrovac is still active. The Agricultural Museum located in Kulpin maintains a well-documented exhibit on old-fashioned ways of growing, retting, braking, spinning, and weaving hemp. The exhibit includes a set of old devices used for domestic hemp processing and hemp folk customs. The book *Praise to Hemp*, by Dr. Jan Kisgeci, details the history of hemp in Yugoslavia.

Today hemp is only a minor crop, grown on several thousand acres for birdseed, certified seed, and fiber products (the former Yugoslavia has five fiber-processing plants and four factories that manufacture rope, twine, and similar hemp-based products for domestic use and export). The only variety presently grown is the local dioecious cultivar "Novosadska konoplja."

The hemp research and development projects coordinated by the Institute of Field and Vegetable Crops in Novi Sad breed new varieties and hybrids. The Institute presently has plans for a new paper factory and expanded textile production.

GLOBAL NETWORKING FOR HEMP

Around the world, renewed interest in industrial hemp has inspired new coalitions and well-attended conferences. In March 1995, the nova Institute of Germany, a leading scientific and environmental consulting organization, sponsored Bioresource Hemp in Frankfurt. The eminent Dr. Ivan Bócsa of Hungary, known as "the grandfather of the hemp field" owing to his fifty years of hemp breeding and research, described Bioresource Hemp as the largest hemp conference ever held. Over three hundred attendees—farmers, government representatives, researchers, entrepreneurs, and environmentalists—gathered for this four-day international conference and trade exposition. Conference

participants networked, presented and discussed research, and organized follow-up meetings around the world. They also ate hemp burg-
ers, relaxed on hemp futons, and viewed prototype hemp plastics,
along with an array of fiberboard, crayons, detergents, shoes, cosmetics, and numerous other products. Attendees left Frankfurt knowing
that an old industry had been reborn. In February–March 1997, the
second Bioresource Hemp symposium was held, again in Frankfurt.
As one sign of the times, the Bank of Montreal sponsored the Commercial Industrial Hemp Symposium in Vancouver, British Columbia,
February 17–19, 1997. On June 1–2 of the same year, the USDA Alternative Agricultural Research and Commercialization Corporation
served as sponsor for the Fiber Futures '97 Conference and Product
Expo, organized by environmental consultant Jeanne Trombley and
the author of this book.

In 1992, in the Netherlands, the International Hemp Association
was born. The IHA publish the outstanding biannual publication *Journal of the International Hemp Association*. In 1995, the Hemp Industry Association was formed in the United States to foster business development among smaller entrepreneurial hemp companies. Also in

*Keynote speaker David Morris addressing
Fiber Futures '97 Conference and Product
Expo in Monterey, California.*

1995, several months after the first Bioresource Hemp conference, the Wisconsin Department of Agriculture held a series of similarly broad-based meetings, organized by Erwin "Bud" Sholts, the agency's director of agricultural development. Out of Wisconsin came the North American Industrial Hemp Council (NAIHC), formed in 1996 with the following mission:

- Form and establish relationships among academia, farmers, agribusiness, manufacturers, marketing firms, government, and public-interest groups, with emphasis on land management, economic, and environmental considerations;
- Develop policies to enhance the stewardship of our lands through the sustainable cultivation, product development, and manufacturing and marketing of industrial hemp and comparable annual fiber crops;
- Promote the development of new products and business based on industrial hemp fiber and seeds;
- Cooperatively foster a better understanding of industrial hemp and other annual fiber crops, and their implications for environmental health and rural economic development.

The NAIHC's founders include Gordon Reichert of Agriculture and Agri-Food Canada; Geof Kime of Canada's Hempline, Inc.; Ken Friedman of American Hemp Mercantile; Jeffery Gain, former executive director of the (U.S.) National Corn Growers Association and current chairman of the USDA's Alternative Agricultural Research and Commercialization Corporation; Andrew Graves, president of the Kentucky Hemp Growers Cooperative Association; Gail Glenn of the Kentucky Governor's Task Force on Hemp and Related Fiber Crops; Senator Lloyd Casey, of the Colorado state legislature; David Morris of the Institute for Local Self-Reliance; John Roulac (this author) of HEMPTECH; and representatives of a Fortune 500 company (International Paper), a $1 billion textile firm (Interface, Inc.), and several environmental groups (including the Earth Island Institute and the Oregon Natural Resource Council).

The fact that such established and credible organizations and individuals now lead the call for a return to industrial hemp has shifted the entire context of this debate in the United States and Canada. Many of the modern hemp industry's early proponents were counterculture types cheering, "Hemp, hemp, hooray!"; they spread the word at a time when no one else seemed to know or even care about hemp. In other countries, the Hemp Wall is falling, including in the United Kingdom (1993), Germany (1996), and Canada (1997). How long will U.S. politicians allow English farmers, German farmers, and Canadian farmers to grow hemp while the hands of their U.S. counterparts are handcuffed?

Now that Fortune 500 executives have started wondering aloud, "Why can't we grow and process hemp for our factories, employ our workers, and do the right environmental thing at the same time?"—hemp's return to North America seems very close at hand.

The nova Institute

The nova Institute, headquartered in Köln (Cologne), Germany, traces its history back to the early 1980s. Ecologically oriented scientists and engineers realized the need to form an independent environmental research institute as a counterweight to the established national research organizations.

Michael Karus, cofounder of nova and coauthor of *Cultivation of Hemp*, recognized that serious research on *Cannabis sativa* was under way, but with minimal communication even among researchers. This led Karus and nova to organize the seminal Bioresource Hemp conferences in Frankfurt. Nova has also been involved in many research, publication, and consulting projects regarding the feasibility of utilizing hemp in a modern economy. The company is widely considered to be the premier hemp-consulting company in the world.

Explains Gero Leson, formerly of nova:

> A major recent accomplishment was the completion of our comprehensive Hemp Product Lines

Project (HPLP). . . . It evaluated which product lines from domestic hemp may be competitive under technical and economic consideration and stand a chance to be implemented within the next three to five years. . . . We found a number of very promising product lines, including specialty pulps, fine textiles from cottonized hemp fiber, and coarse fibers for use in automotive applications. Their implementation may require the installation of innovative processing equipment, or just the willingness of the target industry to experiment with an unfamiliar raw material.

CONSOLIDATED GROWERS AND PROCESSORS INC.

As we go to press, former nova partner Gero Leson has joined with others to form a new company: Consolidated Growers and Processors, Inc. (CGP). This Monterey, California-based company, of which Leson is president and the nova Institute is an advisor, will research, develop, and market industrial crops, including hemp, flax, and kenaf.

CGP is evaluating several initial projects, among which are: an innovative, ecological, and cost-effective pulping technology for fiber plants; the implementation of an advanced process to "cottonize" flax and hemp fibers for use in commodity textiles; the breeding and certification of new hemp varieties with optimized characteristics for specific product lines and climates; and the use of hemp for phytoremediation, which is the efficient clean-up of soils contaminated by heavy metals or radionuclides.

Rohemp

Karl F. Ströml's interest in hemp was sparked when Frank Waskow of Katalyse Institut in Köln (Cologne), Germany, told Ströml that Romania is one of the traditional hemp-growing countries of the world. Ströml decided to found a company there, and named it Rohemp ("Ro" for Romania).

After two years spent living and working in Romania, Ströml started another Rohemp company in his native Austria. He is currently founder, major stockholder, and president of Rohemp S.A., Bucharest, and also the executive manager of the Austrian company, Rohemp GmbH. Most of Rohemp's work is in industrial-hemp research, but the company also makes approximately sixty products, including fabrics, yarns, paper, books, ropes, hats, caps, and bags.

Ströml is quite enthusiastic about Rohemp's recent accomplishments:

> [In 1996,] we developed and built a mobile hemp braker—a self-driven decortication machine. The low-quality fibers you can use for the construction or insulation industry; the hurds are used here to make paper. We have developed together with a big Austrian paper-pulp factory the first hemp paper made from hurds and the first hemp paper made TCF (totally chlorine-free). We sell it under the label Rohemp Legalise.

Rohemp GmbH
Wall Strasse 36
A-8280 Fürstenfeld, Austria
Telephone: +43 (0) 338252300
Fax: 52301 or 5230015
e-mail: rohemp@styria.com
www.hempworld.com/rohemp (English)
www.hanfnet.de/rohemp.html (German)

Columbian Twine

CATALOG No. 58

MANUFACTURED BY

COLUMBIAN ROPE COMPANY

AUBURN, "The Cordage City," NEW YORK

NEW YORK, N. Y.
171 JOHN ST.

BOSTON, MASS.
38 COMMERCIAL WHARF

CHICAGO, ILL.
2820 NORTH PULASKI ROAD

NEW ORLEANS, LA.
4400 FLORIDA AVE.

The Columbian Rope Factory in New York state manufactured U. S.-grown hemp for twine and rope up until the late 1950s.

I believe hemp is going to be the fiber of choice in both the home furnishings and fashion industries and I wanted to be one of the first designers to use this fabric in an innovative way. I love the fabric. The plan is for us to eventually introduce 100-percent hemp products.

CALVIN KLEIN

Hemp in the Marketplace

CHAPTER FIVE

TODAY'S PHENOMENAL global resurgence of industrial hemp bears witness to the powerful commercial potential of this crop. As the accompanying graph shows, worldwide gross retail sales of hemp products are accelerating steadily each year since a veritable explosion of 1,000 percent between 1993 and 1995. The 1996 gross retail sales of hemp-derived products in the United States are estimated to have exceeded $40 million; worldwide, the figure is approximately $75 million (not including mainland China).

Despite widespread domestic demand for hemp-based products, American farmers still are prohibited from raising this hardy, versatile crop. Raw or finished hemp materials must be imported, funneling most profits to other countries and increasing our nation's trade deficit while infringing upon the economic rights of growers, manufacturers, and entrepreneurs. The official exclusion of industrial hemp from the federal government-approved list of fiber resources impedes buyers and sellers from using the raw goods of their choice. This is tantamount to a socialist or communist policy whereby the state, at its sole discretion, determines which fiber resources can compete and which are prohibited.

97

Now imagine that a valuable commodity such as cotton or timber were for some reason illegal to grow and process in one's own country. Businesses would be forced to sit on the sidelines as foreign competitors dominated the world market for thousands of products in major industry segments. Wouldn't the commercial sector raise a hue and cry for repeal if laws prohibited it from engaging in legitimate commerce that other nations were profiting from? Wouldn't ingenious entrepreneurs try to find ways to work within the legal parameters to profit from the outlawed commodity? This is the situation for industrial hemp today in the United States.

HEMP BUSINESS STRUCTURES— COTTAGE TO CORPORATE

In 1993, a new hemp business was started approximately every month in the United States, but in 1996, several new ones were being formed every week. As in many expanding commercial sectors, hemp enterprises run the gamut from home-based industries to multinational corporations.

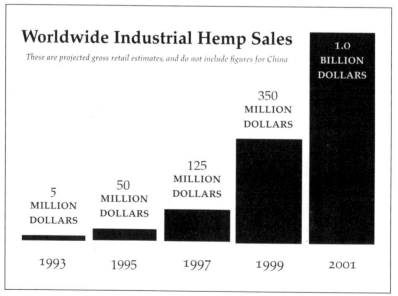

Worldwide Industrial Hemp Sales

These are projected gross retail estimates, and do not include figures for China

5 MILLION DOLLARS	50 MILLION DOLLARS	125 MILLION DOLLARS	350 MILLION DOLLARS	1.0 BILLION DOLLARS
1993	1995	1997	1999	2001

© 1995 HEMPTECH

Cottage Industries

While cottage industries are certainly entrepreneurial structures, they constitute a unique subset of commercial hemp activities. Many people are not interested in building a fast-growing, high-pressure business; they would rather be the owner/artisan/chief bottle-washer who make unique, value-added products at home or in the studio. The raw materials of hemp, ranging from fiber to seed to oil, offer talented and creative individuals a host of income-producing opportunities. Hand-crafted textile arts, hand-crafted clothing, cosmetics, foodstuffs, and furniture are only a few of the hempen products offered by cottage industries worldwide.

Increasing numbers of workers have been developing home-based businesses as an alternative to dreary, unfulfilling, and/or high-stress jobs. Cottage industries provide a wonderful avenue for families that want to raise their children the old-fashioned way—living and working as a close-knit team. Moreover, many men and woman are quite satisfied with a simpler lifestyle, knowing that they are contributing to a shift in global priorities from consumption to sustainability. The migration of Americans back to rural areas, combined with the digital revolution, will form the genesis of a new wave of home-based businesses in the coming decade.

The primary challenge for any cottage industry is to keep up—in the home environment—with the many tasks required to run any business successfully, such as accounting, marketing, manufacturing, and shipping. A professional Web site, together with virtual support services for distribution and marketing, can give cottage industrialists a leg up in creating successful hemp enterprises.

Carol Miller of Cazadero, California, runs Hemp Essentials, a thriving home-based business in hemp and herbal soaps, lip balms, and other bodycare products. Her family participates, growing herbs as well as helping her blend, package, and ship orders. By carefully managing her production and storage capacity, Miller has been able to keep eight part-time jobs going for herself and family members.

Entrepreneurs

A good entrepreneurial firm, whether a startup or a thirty-year-old family enterprise, responds quickly to new customer interests and readily spots emerging, unusual trends. The computer industry is re-plete with examples of small, agile companies beating out corporate giants in the anticipation and introduction of innovative technologies. Such is also the case in the hemp industry. At the 1996 New York Yarn Fair, the nation's biggest yarn manufacturers were caught by surprise when the hottest new products included the first American-spun hemp yarns to be produced in fifty years—yarns released by several small firms in partnership with the modest-sized Hemp Textiles International of Bellingham, Washington.

Small startup enterprises are rapidly staking claims for a piece of the hemp industry. Well over five hundred American hemp companies do $50,000 to $500,000 in annual sales. At least a dozen more, such as Hempstead, Two Star Dog, Ecolution, Ohio Hempery, and American Hemp Mercantile, annually surpass $1 million. These successful entrepreneurs contribute economic vitality to their local communities and are indicative of hemp's breathtaking commercial momentum.

Because of the U.S. government's current hemp prohibition, such firms are importing hemp fiber, oil, or seed from Europe, China, and elsewhere. The majority of these companies manufacture twine, cloth-ing, accessories, foods, and bodycare products, to be sold at retail out-lets around the country and also through virtual hemp-product stores on the World Wide Web.

Yet there is always a danger for companies that offer just another "me too" hemp product or line. As the market matures, simply offer-ing hemp-based products will not be enough to guarantee a reliable stream of customers. Successful firms must offer superior design, qual-ity, customer service, and environmental standards, combined with competitive pricing and well-managed marketing, distribution, and fi-nancial management systems.

Multinational Corporations

In this era of rising populations and scarcer resources, major corpora-tions with global operations are paying increasing attention to the se-curing of dependable, long-term supplies of raw materials. At the same time, corporate managers have begun to think in new ways, as re-flected by the environmental and sustainability standards set by the Swiss-based International Organization for Standardization (ISO), along with the European Union's requirements for manufacturers to build recyclability into products ranging from packaging materials to automobiles.

"ISO 14000" is the ISO's generic series of environmental management standards, devised to blend or harmonize ecological requirements for companies around the world, with the goal of giving corporations a framework for managing their environmental impacts. The United States Department of Energy is now requiring all contractors to register as complying with the new standards. ISO 14000 will enable other federal government agencies, along with corporations and consumers, to know that a company, product, or process meets international expectations for environmental quality—without having to do all the detective work themselves.

Monsanto Corporation, which gave the world Roundup, Butachlor, and bovine growth hormone (BGH), is one of pollution's villains, but the company is attempting to clean up its act for the twenty-first century. According to Monsanto CEO Robert B. Shapiro, in an interview in the January–February 1997 *Harvard Business Review*:

> Sustainability involves the laws of nature—physics, chemistry, and biology—and the recognition that the world is a closed system. What we thought was boundless has limits, and we're beginning to hit them. That's going to change a lot of today's fundamental economics, it's going to change prices, and it's going to change what's socially acceptable. . . . If emerging economies have to relive the entire industrial revolution with all its waste, its energy use, and its pollution, I think it's all over.

Industrial hemp offers large corporations a dependable raw material: It matures in one growing season and is easily cultivated in many climates. Hemp also emanates an appealing aura of environmental sensitivity, since it grows robustly without herbicides or insecticides, and can replace many products now manufactured from cotton, timber, or petrochemicals.

The German Aerospace Institute conducted research in 1994 and 1995 revealing that twenty-five pounds or more of hemp fiber per car could be used in manufacturing automobiles—for carpets, gaskets, ceiling liners, seat covers, floor mats, and interior paneling. In 1996, German automaker BMW introduced a car that uses hemp fiber as part of its air-bag system. Daimler-Benz is also testing hemp fiber for interior paneling. Yet German hemp farmers complain that these giant firms want to purchase their harvest at a very low price. For now, synthetics still cost less then hemp.

In the United States, two Fortune 500 companies, International Paper, the world's largest paper company, and Inland Container, have been studying the use of hemp fiber since 1996. The American Farm Bureau, the Wisconsin Agribusiness Council, and the Canadian Auto Workers Union all have endorsed test planting and research.

Hemp Makes the Business Page

On January 5, 1997, the *Washington Post* ran an article about hemp, by John Mintz, entitled "Splendor in the Grass?" Not only did this feature appear in the business section instead of in the main section, where hemp stories usually have been placed, but it was a lead article in the Sunday edition. The article explained: "The hemp plant comes in two varieties. One, its proponents say, could transform industries and provide an environmentally safe source of wonder products. The other is marijuana, and therein lies a debate that has business warily stalking a tantalizing raw material." Mintz went on to quote Curtis Koster, technology business manager of International Paper: "The company is intrigued by hemp as a way to address what timber interests call the looming worldwide fiber crisis. . . . It's the strongest, easiest to grow, and has the broadest geographical range. . . . There's no doubt excellent paper can be made from hemp. . . . It is a remarkable thing God put on earth."

The article also pointed out: "The need for paper and other fiber products (such as fiberboard, packing materials, and pulp) is skyrocketing in the Third World—it seems demand for such products precisely tracks economic and educational progress."

Interface, Inc., a U.S.-based multinational textile firm, has been developing a sustainable materials policy. In late 1995, I made a presentation to their executive team on the potential offered by hemp fiber. Interface has since identified hemp as a promising material for their carpet-tile products, sold for commercial use around the world. In Thomas Petzinger, Jr.'s July 11, 1997, *Wall Street Journal* column, "Business Achieves Greatest Efficiency When at Its Greenest," Interface was mentioned as substituting certain yarns with hemp and flax, ". . . a step toward carpeting that is both 'harvestable' and compostable."

Ian Low of the U.K.'s Hemcore has observed that "after we have spent lots of time and resources in solving hemp's challenges, the big players may well push smaller firms aside." As the hemp industry matures, smaller companies will need to select their niches carefully. One way to distinguish a given product, so that it is not identified as "just another hemp commodity," is for entrepreneurs and cottage industrialists to work with the organic agriculture industry to develop goods that are certified organically grown. For instance, the Organic Fiber Council, based in Davis, California, serves as a trade association for farmers and manufacturers.

CHALLENGES TO GROWTH IN THE HEMP INDUSTRY

Given that hemp has been *materia non grata* in much of the Western world for more than fifty years, various roadblocks stand in the way of the industry's resurrection. First and foremost, of course, is the de facto ban on hemp cultivation in the United States, as discussed in chapter 3. Another problem is the low supply of hemp relative to worldwide demand; this issue relates both to hemp's restricted legal status and to certain technological obstacles.

Inefficient equipment—including machinery for harvesting fiber and seed, and for fiber separation—requires substantial improvement before hemp can move beyond its current role as a small but exciting niche crop. Research on the production of high-yield seed and fiber cultivars, and on a more cost-effective alternative to traditional field retting, is also essential. The remaining chapters of this book deal with these agricultural matters more thoroughly.

From a primarily commercial perspective, significant challenges arise regarding processing centers, European subsidies, sustainable manufacturing, and marketing.

Processing Centers

Some people naively assume that farmers who plant hemp will immediately get rich because there are so many eager buyers for hemp stalks. Yet, unless these farmers have access to regional processing centers, few buyers may exist. By way of example, let's explore the experiences of three prototypical European hemp processors.

/ HEMCORE / In 1993, Hemcore of Essex, England, contracted with U.K. farmers to grow several hundred acres of hemp. The company then sold the core fibers of the resulting hemp stalks as animal bedding, but there were as yet few buyers for the more valuable long fibers. To develop the latter market, the firm spent several years educating clients; developing joint-venture research projects; and supplying sample fibers to various paper, textile, construction, and automotive companies. With Hemcore's modern processing facility, and with these years of preliminary work behind it, the company has been able to approach a total planted acreage of six thousand acres in 1997.

/ BADISCHE FASERAUFBEREITUNG / Upon the lifting of Germany's ban on hemp in early 1996, German farmers planted three thousand acres that spring. As discussed at the nova Institute's 1997 Bioresource Hemp conference in Frankfurt, the maximum optimal distance from field to processing center is about thirty miles; it is uneconomical to ship bulky hemp stalks much farther in Europe, where fuel costs are higher in the United States. The Badische Faseraufbereitung company successfully rose to this challenge and jump-started a processing facility when the ink was barely dry on the documents ending the German hemp ban. This firm began by contracting with farmers to plant 250 acres (to yield approximately one thousand tons), yet has built a facility that could eventually process up to ten thousand tons of hemp stalks per year. Much of this long, coarse fiber will be used in the automotive industry for such purposes as interior panels. The core fibers will be sold to the animal bedding market and to specialized manufacturers of construction materials.

HempFlax

HempFlax's processing facility in Oude Pekela, Netherlands.

/ HEMPFLAX / In Oude Pekela, the Netherlands, HempFlax planted its first three hundred acres in 1994. Like Hemcore, the Dutch company has since paid its dues and built a processing center at its location near the German border. By 1997, HempFlax was contracting with Dutch and German farmers to grow approximately four thousand acres, and paying the farmers $80 to $100 per ton (over and above the subsidies paid them by the European Union). Since its inception, the company has warehoused much of its processed long fiber in anticipation of expanded local markets. In fact, HempFlax plans to operate a hemp-pulp mill adjacent to its processing center in Oude Pekela by late 1998.

European Subsidies

The European Union (EU) provides direct subsidies to farmers in member nations for growing a variety of agricultural crops, including up to $400 per acre for nonfood commodities such as hemp and flax. This subsidization has made it possible for European hemp companies to build the processing infrastructure necessary for commercial success.

It is highly unlikely that the governments of the United States, Canada, or Australia will ever pay such subsidies; this will create challenges for farmers and entrepreneurs in these countries as they work to develop their own profitable hemp industries. However, American,

Short- to Medium-Term Market Potential for German Hemp

Product Line	Intermediate Products (metric tons/year)	Farming Area (hectares/year)	Market Value of Intermediate Products (million DM/year)
Bast Fiber			
1. Specialty pulps for technical applications	~ 5,000 well-decorticated fiber	~ 3,500	~ 4
2. Press-molded interior panels for automotive applications	~ 4,000 mechanically processed coarse fiber	~ 2,500	~ 4.4
3. Geotextiles for erosion control, etc.	~ 4,000 mechanically processed coarse/medium fiber	~ 3,000	~ 6
4. Needle-punched carpeting	~ 3,500 mechanically processed coarse/medium fiber	~ 2,500	~ 5.3
5. Textiles for clothing; cottonized hemp as a cotton substitute	~ 15,000 physically-chemically processed fine fiber	~ 13,000	~ 45
6. Mats for thermal insulation in construction	~ 10,000 physically-chemically processed fine fiber	~ 8,000	~ 30
Total of Bast Fiber Product Lines		~ 30,000	~ 95
Hurd			
7. Animal bedding	~ 90,000	~ 30,000 (combined production)	~ 18
Seeds and Oil			
8. Foods	~ 1,200–6,000 seeds or ~ 300–1,500 oil	10,000 (combined production) or 1,000 (seed production only)	~ 5.2–25 (50% seeds/ 50% seeds as oil)
9. Natural cosmetics	included in "Foods"		
10. GLA for pharmaceutical and cosmetics industry	150 oil	500–1,000	~ 3.8

Source: nova Institute, 1996

Canadian, and Australian farmers do manage to successfully compete in world grain markets, despite the lesser subsidization for their crops. Additionally, the long-term trends of rising fiber prices and the societal switch toward more sustainable products may well create more favorable economic conditions for hemp.

Sustainable Manufacturing

As explained later in this chapter and throughout this book, industrial hemp can make positive contributions toward the development of sustainable products. As the industry progresses, there will be increasing interest and questions regarding how hemp products are manufactured. For example, what type of dyes will be used in making textiles, or glues in building fiberboard products? Will industry merely substitute hemp for timber and cotton—without a thorough evaluation of established, perhaps polluting, processes?

Because hemp is such a new albeit old material, companies have a unique opportunity to research, test, and implement leading-edge technological processes that improve upon existing environmental standards. I anticipate that there will be a tremendous opportunity to develop green product labeling standards, which allow consumers to identify products and corporations that care about our natural resources. Some industry participants may use hemp as a vehicle to leap-frog past antiquated manufacturing practices and develop innovative, Earth-friendly hemp products.

Marketing

Hemp's reputation as a forbidden fruit of the counterculture presents both a boon and a challenge to marketers. While some firms successfully capitalize on hemp's taboo aura, others are marketing hemp as a sustainable, renewable fiber with a long, respectable history. Given the prevalence of confusion and misinformation about industrial hemp, consumer education will remain a vital component of hemp-based product sales for the near future. Of course, as sales increase, the need for explaining that hemp is not marijuana will gradually decrease.

Those entrepreneurs working to deal with hemp's counterculture image may have to tread an uphill road for the next several years, but hemp's many strengths are a marketing executive's dream come true:

- Hemp is longer-lasting and stronger than wood or cotton;
- Hemp has antimildew properties;
- Hempen clothes are warmer in winter and cooler in summer;
- Hemp seed and oil have the most ideal ratio of essential fatty acids of any edible plant;
- Hemp conserves forests;
- Hemp provides an alternative to petrochemical products;
- Hemp can be farmed without toxic chemicals;
- Hemp supports family farmers and small businesses;
- Hemp manufacturing is less polluting;
- Hemp is one of the oldest cultivated plants in history and goes back at least six thousand years;
- The great Renaissance artists painted on hemp canvas; the word "canvas" itself is derived from the word "cannabis";
- George Washington and Thomas Jefferson grew hemp; Abraham Lincoln also supported its cultivation in the United States.

HEMP AND BETTER BUSINESS PRACTICE

By utilizing a renewable resource, the hemp industry has the potential to help move our economy in a more sustainable direction, and one that is more socially and environmentally responsible. Since hemp grows well in most climates and offers amazing product versatility, it is perfectly suited to bioregional economics, which is defined by natural land boundaries (such as a valley or a river basin) and seeks to stimulate commercial activity that benefits both people and the ecology. Producing goods for local use benefits the local job and tax base, to a far greater extent than importing the same goods from thousands of miles away.

The True Cost of Goods

Before the advent of modern advertising, there were few national brands for common household items such as candles, clothing, or cleaning products. Most goods were made from plant-based materials that often were grown and processed in the same region where they were sold.

Today, the world's best-selling nonfood brands are made primarily from petrochemicals and/or rely on petrochemical additives. Manufacturing and distribution systems (shipping, for example) depend upon fossil fuels, often transported over great distances. Modern products typically are packaged in brightly colored, resource-intensive materials—paper and plastic—and sold by means of expensive marketing campaigns designed to build brand recognition and loyalty. Additional corporate overhead accrues through high executive salaries and bonuses.

Because the industrial processes that yield many major brand-name products rely upon deforestation and the use of toxic petroleum byproducts, significant environmental liabilities and public-relations nightmares occur. This risk requires corporations to retain expensive law firms for defensive purposes and liability insurance policies; to hire entire departments to create prefabricated media sound-bites; and to give large donations to politicians in order to maintain corporate subsidies and/or weaken environmental regulations. Thus, the price we pay at the retail level for the actual product—be it a bar of soap or a finished garment—accounts for a very small percentage of its true cost.

It behooves us to create a more level playing field in the marketplace, in order to give corporations and entrepreneurs a fairer shot at building businesses around better-quality products (such as hemp), improved environmental performance, and enhanced employee working conditions. However, our current situation is such that the largest polluters, from agribusinesses to utilities to oil and timber companies, receive enormous government subsidies that keep their bloated corporate ships afloat. For all the talk of a free-market economy, many large corporations operate under a socialist corporate model rather than a true capitalist model. As shocking as it may seem, the federal government's most expensive support program is not welfare for single mothers, but welfare for corporations. Conservative estimates of corporate welfare to American industries range from several hundred billion dollars to upwards of a trillion dollars in direct and indirect annual subsidies, including the following:

- Storing nuclear waste from corporate utilities at taxpayer expense;

- Leasing public lands below cost for cattle grazing;
- Selling mining rights on public lands at less than a penny on the dollar;
- Building expensive roads in national forests for money-losing logging operations.

The preceding examples of corporate welfare apply in some measure to numerous other countries, including Japan, China, Canada, France, Germany, the United Kingdom, and Australia.

A New Paradigm

On the other hand, many of the individuals working in large corporations are good people who are doing the best they can to improve their firms' operations. In fact, a renaissance of leading-edge thinking on environmental and social policies has begun to sweep major corporations around the world. The United Nations–sponsored "Zero Emissions" initiative, for example, is working with many Japanese and European firms to eliminate the release of hazardous emissions in manufacturing. The Natural Step program, which originated in Sweden, educates entire companies to reevaluate how products are made and ultimately consumed back into our environment.

This call for a responsible corporate ethic is compatible with the growth of the hemp industry and the expansion in human and environmental concerns at large. Although we are dealing with a systemic condition that will not be easy to change, there are signs of hope.

The way we do business *is* changing. Tens of thousands of entrepreneurs and business executives have seen the debilitating results of polluted rivers, deforested hillsides, and toxin-laden neighborhoods in communities around the world. Instead of joining another environmental group to try to lobby already bought-and-paid-for politicians, businesspeople are using commerce as a tool with which to create a cleaner, more healthful environment.

Ray Anderson, chairman of Interface, Inc., refers to himself as "an industrial ecologist" and has given his textile corporation a mandate to achieve sustainability goals within its operations. Meanwhile, in New York City, the recycling group Bronx 2000 is transforming landfill-destined broken pallets into valuable resources by remanufacturing them for furniture, flooring, and other products. And consider the many garment-cleaning firms that are substituting citrus-based wet-cleaning processes for the traditional dry-cleaning concoction of petrochemical-based Perc, which contains chemicals known to be hazardous to workers and customers alike.

Imagine the effects of diverting the flow of hundreds of billions in annual sales from wasteful, polluting, centralized, petroleum-based industries to sustainable, regionally appropriate, carbohydrate-based industries. A convergence of new materials, new technologies, and new perspectives amounts to a new paradigm for conducting business—one that affirms the importance of human and planetary well-being.

Growing hemp instead of chemical-intensive cotton for textiles is one of many signposts that "the business of business" is changing dramatically. Consumers who patronize ecologically aware companies are voting with their dollars. In fact, society is changing much faster through "dollar voting" than it can through ballot-box voting for status-quo politician A or status-quo politician B. The bottom line is that businesses that refuse to change risk losing market share.

A New Household Word

In the coming years, an increasing number of increasingly aware consumers will purchase tens of billions of dollars' worth of clothes, homes, automobiles, and other goods. Hemp is emerging as a powerful marketing icon for millions of savvy buyers around the world. Some of the world's most successful consumer brands already have welcomed it as an exciting new material: Adidas sells a hemp shoe, Calvin Klein uses hemp for bedroom accessories, and Giorgio Armani has designed a full line of hempen men's and women's clothing.

Hemp's strongest North American supporters are in the fourteen-to-twenty-five-year-old age group, especially in the trendsetting West Coast cities. This generation will become the major consumers of the next several decades. In the year 2003, new-car ads may well tout "luxuriously upholstered hemp seats," or a jeans company may invite buyers to "slip into soft hemp jeans that will last a lifetime."

The word "hemp" is becoming a powerful emotional symbol of quality, durability, practicality, Earth-friendliness, and style—all wrapped in an aura of counterculture appropriateness for our postmodern era.

So, while Toyota, Levi's, and IKEA will remain strong brand names, consumers will respond positively to companies that incorporate the word "hemp" in their marketing.

After the passage of Vermont's 1996 hemp bill (see chapter 3), a study conducted by the University of Vermont assessed citizen interest in purchasing hemp products. A random sample of 770 Vermonters (average age forty-seven) was conducted by telephone. The survey results clearly showed that citizens of Vermont are ready and willing to purchase hemp products:

- If hemp jeans were priced competitively with cotton jeans, 54 percent would substitute hemp for cotton in all current purchases;
- 37 percent would pay more for hemp jeans than for cotton ones;
- 69 percent would buy hemp-based computer paper, if it were price-competitive;
- 67 percent would pay between 2.5 and 10 percent more for hemp-based writing paper.

Regional manufacturers of hemp products from soaps to jeans may capture market share from entrenched corporate brands by using guerrilla-style marketing tactics. In progressive communities, the catchphrase "Grown, manufactured, and marketed in Hometown, U.S.A." will attract sales, even if such products are priced a bit higher than national brands.

As long as we are engaged in a bit of forecasting, which is as much an art as a science, let's throw in the wild card of doubled oil prices over the next decade or two. Transportation costs for petrochemically derived goods from centralized factories may also skyrocket.

Is this a likely scenario? Who knows for sure? Corporations and entrepreneurs should certainly be prepared to hedge their bets, or—better yet—to profit from such an outcome. Companies that wait until all the evidence is in may well miss the opportunity. Whether an artist, an investor, an entrepreneur, or an executive, the individual contemplating a hemp business is responsible for conducting the research and making the assessments that will determine its commercial feasibility.

JOBS AND BUSINESS OPPORTUNITIES

Throughout the business world, corporations continue to downsize and to restructure themselves. Displaced workers seek new jobs and business opportunities, often closer to home or on a smaller scale. Today's top-selling management books promote ethical and spiritual business practices along with "corporate greening." As more consumers look for high-quality, long-lasting goods manufactured in an environmentally responsible manner, the demand for hemp products is rising.

Industrial hemp thus offers significant business and employment opportunities in such broad commercial sectors as farming, processing, manufacturing, retailing, recycling, and cooperatives. Rural economies in particular will receive a boost once American hemp manufacturers and processors are allowed to reduce transportation costs by locating their operations close to hemp-growing regions in this country. Certainly the number of new jobs in any given location will grow as the hemp industry grows. For example, Two Star Dog, the hemp-based apparel firm located in Berkeley, California, has added twenty positions in the last three years.

Hemp entrepreneurs would do well to remember the lesson of the 1849 gold rush: While few miners made fortunes, many merchants did —by selling picks, shovels, Levi's, and provisions to the miners. Those businesspeople who strike while the iron is hot will be amply rewarded. Consider the following examples of potentially profitable hemp business opportunities:

- A Banana Republic–style hemp retail chain;
- A Hard Rock Cafe–style theme-based hemp restaurant;
- A manufacturer of processing equipment for hemp seed or seed oil for home and commercial use;
- A manufacturer of equipment that solves fiber-processing bottlenecks;
- A virtual-marketing-distribution organization that unites thousands of smaller firms to gain more profitable market access;
- Regional representatives for manufacturing firms that specialize in particular niches, such as processing equipment, consumer products, and so on;

- Allied support services that specialize in the industrial hemp sector, such as accounting, advertising, and business management.

Individuals looking for jobs in the hemp industry should learn as much as they can about the field, then contact some of its fast-rising companies.

Profiles of several of today's leading hemp firms appear throughout this book (a more complete list is in the appendix). Each business is unique, yet all are on the leading edge of a commercial development, with the potential to have major economic and environmental impact in the twenty-first century.

Investing in the Hemp Industry

Much as Dustin Hoffman's character in the 1960s movie *The Graduate* was told that "plastic is the future," the sharp investment banker of the late 1990s is telling young men and women that industrial hemp is the future, for it offers a unique, sturdy, and versatile material for thousands of consumer and industrial uses. Only time will tell whether hemp can move from the status of a promising minor crop to one rivaling corn or cotton.

But people are increasingly aware of hemp's commercial possibilities. One of the questions most frequently asked of me when I talk on the radio or make public presentations is "How can I invest in this profitable industry?"

As this book goes to press, no publicly traded, predominantly hemp-based firms are listed on the U.S. stock market, but it is likely that several companies will go public in the coming years. Does this mean that they will be surefire investments? Not necessarily: as in any fast-paced industry, some ventures will succeed and some will fall by the wayside. Because raw hemp materials remain so expensive, many outfits are struggling merely to stay in business.

Nevertheless, some companies will surely develop very solid and profitable businesses. One can look at Whole Foods, the leading retailer in the natural-foods industry, to see how handsomely investors have been rewarded from stock purchased over the last five years.

Today's young hemp companies are looking for high-net-worth individuals who can invest a hundred thousand to a million dollars.

Obviously there is significant risk, with no guarantee of return on investments. When evaluating any enterprise, including any hemp business, potential investors should consider the caliber of the company's management team, the soundness of its business strategy, and the company's competitive advantage in its market niche.

HEMP PRODUCT VERSATILITY

In 1938, the *Popular Mechanics* article "New Billion-Dollar Crop" reported that "hemp . . . can be used to produce more than twenty-five thousand products, ranging from dynamite to cellophane." As you read this quote, consumers from North America to Europe to Australia are purchasing hundreds of different products made from hemp. Hemp jeans, shirts, and hats are becoming fashionable from Hamburg to Los Angeles to Tokyo.

Hemp's positive commercial impact will be felt in these major world industries: agriculture, automobiles, body care, construction materials, feed, food, furniture, industrial resins, paper, plastics, and textiles. Here is a closer look at some of hemp's numerous practical products.

Animal Care

Hemp seed has long been popular as a feed stock for a variety of animals. Birds are especially attracted to hemp seed for its superior nutritional qualities. To this day, most birdseed mixes contain hemp seed, for professional breeders swear by the benefits of including it in the diets of their birds. After hemp seed is crushed to extract the oil, the remaining seed cake is approximately 25 percent protein, and makes an excellent feed for pets as well as for cows and chickens.

Hemp's short fibers make an ideal bedding for horses or for such small, caged animals as hamsters and rabbits. It also serves as an excellent cat litter.

La Chanvrière de l'Aube

Super-absorbent hemp animal bedding.

Hemp has twice the moisture-holding capabilities of straw or wood shavings. It also lasts longer, is more hygienic, and composts faster. No wonder the horse paddocks at Buckingham Palace are refilled with hemp core fiber on a regular basis.

Automobiles

As noted earlier in this chapter, recent research by the German Aerospace Institute has shown hemp's suitability for making auto components such as gaskets, seat covers, floor mats, and interior paneling. (BMW already has used hemp fiber in the air-bag system for one of its 1996 Series 7 models.)

Several American firms are researching the use of hemp fiber as an economical replacement for fiberglass in electric vehicles made of lightweight composites. Scientists at Daimler-Benz have already experimented with a variety of natural alternatives, such as flax, to avoid using environmentally unsound glass fiber. (The problem with glass fiber is that it cannot be recycled without breaking into small particles that are dangerous if inhaled.) A statement from Daimler-Benz notes:

> Hemp fibers have a number of advantages over flax. They are more rich than flax and can be cultivated without the use of insecticides. Initial investigations have shown that hemp matches and even surpasses flax in terms of performance potential and promises to be more economical.

Body Care

Numerous personal care products can be manufactured using the oil extracted from hemp seeds. Research has shown that hemp oil assists the body's natural ability to heal, both externally and internally: Its essential fatty acids are readily absorbed into skin cells. Because hemp oil can help restore and moisten skin, it is becoming popular for use as a massage oil and in lip balms, soaps, shampoos, and lotions. When combined with herbs, hemp-oil salve assists in healing skin irritations, insect bites, and minor cuts.

In Europe, the essential oil is being extracted from low-THC hemp leaves for use in perfumes. In the field of medical aromatherapy, this oil is also being applied topically for the soothing of sore muscles.

Construction Materials

Fiber composites, the fastest-growing segment of the wood-products industry, comprise the largest potential market for industrial hemp.

Composites include paneling, medium-density fiberboard, plywood trusses, and support beams. At current timber-harvesting levels, some composite mills will need to find an alternative fiber source in order to stay in business. These factories can substitute hemp for wood without changing existing production equipment. Washington State University's preeminent Wood Composite Laboratory has tested hemp for use in medium-density fiberboard, and the results show that hemp is twice as strong as wood. According to lab director Tom Maloney, "The use of hemp fiber in multidensity fiberboard and other composites looks very promising." Masonite Corporation, a division of International Paper, is researching the use of hemp fiber in manufacturing building products.

Hemp also can be processed into a variety of insulation products that are safer than fiberglass and easy to install. A French firm, La Chanvrière de l'Aube, has been converting hemp hurds into a fluffy

La Chanvrière de l'Aube

French house utilizing hemp plaster on outside walls, and hemp-fiber insulation in interior walls and attic.

cellulose product that is blown into walls, attics, and air spaces, or placed there in bags. The builders of more than a thousand French homes have utilized a cement-plaster–like material made from a combination of hemp hurds and lime. The material can be used without further additives in foundations, walls, floors, and ceilings, and for interior or exterior plaster. It is stronger than concrete, yet five times lighter, and has excellent insulation and fire-retardant properties. It is also resistant to insects and mold. Of the French companies marketing this material, Chenovette Habitat and La Chanvrière de l'Aube are two leading ones. The material has also been used to remodel several German houses and to build a new home in Quebec.

Foods

Hemp seed also is a nutritious food source for people: high in protein, calcium, magnesium, phosphorus, potassium, and vitamin A. Like the soybean, hemp seed can be made into many different food products, including nondairy cheese, milk, and ice cream; yet it is easier to digest. Try mixing hemp seeds, nuts, and honey for a nourishing confection. Udo Erasmus, Ph.D., the author of *Fats that Heal, Fats that Kill: The Complete Guide to Fats, Oils, Cholesterol, and Human Health*, states that "hemp butter puts peanut butter to shame for nutritional value." Hemp seeds can be ground, soaked, or crushed for their oil content. After the oil has been extracted, the remaining meal or seed cake makes a valuable nutritional supplement in high-quality breads, pastas, cakes, and cookies. (The seed cake has a short shelf life and thus should be used in fresh-baked products.)

Paralleling many people's strong interest in hempen foods is a nutritional focus on adding essential fatty acids (EFA's) to their diet. Although the body needs these "good" fats for healthy skin, hair, eyes, and overall physical health, it does not produce them. EFA's come from only a few food sources, such as fish oil and the seed and oil of hemp, flax, borage, or primrose. At a volume level of 81 percent, hemp oil is the richest known source of polyunsaturated EFA's. It also contains gama linoleic acid (GLA), a very rare nutrient. Hemp oil can be taken as a nutritional supplement similar to flax-seed oil, or can be used in salad dressings and other oil-based recipes. (Like other fragile oils, it should be refrigerated to prevent rancidity.)

Hemp-seed oil certainly offers promising markets for nutritional supplements. By comparison, organic flaxseed oil now has a multi-million-dollar market, yet hemp has a better taste, a longer shelf life, and a more balanced EFA profile than flax.

Fuels

Technically, hemp makes an excellent fuel source. Its short fibers can be burned for heat production in factories or converted into methanol fuel for vehicles. (You will recall from chapter 1 that Rudolf Diesel designed his namesake engine to run on vegetable oils such as soy, hemp, flax, etc.) Yet, realistically speaking, many more economical sources of plant fibers and vegetable oils exist—not to mention petroleum. Hemp fiber and oil will be used for other, higher–value-added products such as those described in this chapter. Perhaps, over the long run, hemp fuels will become more economically feasible compared to other fuel sources.

Paints and Sealants

Until the 1930s, linseed and hemp oils made up the majority of all resins, paints, shellacs, and varnishes. In Texas, for example, the Sherwin-Williams paint company used hemp oil in its products. Yet, with the advent of cheaper petroleum-based compounds, combined with the Marihuana Tax Act of 1937, hemp products disappeared from the marketplace.

Paper

In 1995, I published a small book printed on tree-free hemp paper. Hemp has a fiber-yield-per-acre several times higher than that of trees. Hemp's long and tough bast fibers, while requiring cutting prior to paper-making, can produce high-quality papers for books, magazines, and stationery. The shorter core fibers, blended with another long-fibered pulp, can be used to make newspaper, tissue, and packaging materials. Until the late nineteenth century, the world relied on annual crops such as hemp, flax, and cotton for paper-making.

Hemp fiber's low lignin content accommodates environmentally benign bleaching, without the use of harsh chlorine compounds. Such tree-free paper mills, unlike some wood pulp mills, would sustain healthy fish populations downstream. Hemp paper resists decomposition and does not yellow with age, in contrast with wood-derived paper. In fact, hemp paper more than fifteen hundred years old has recently been found. Furthermore, because hemp fiber is so strong, hemp paper can be recycled several times more than paper made of wood.

Rising demand, combined with accelerating deforestation, is motivating the paper industry to explore nonwood fiber sources such as hemp. British, Canadian, and German mills have begun to produce hemp paper, and Kimberly-Clark, an American company, produces hemp paper in France for Bibles and cigarettes. Emerging technologies will permit more cost-efficient pulping of hemp fibers. According to Dr. Med Byrd, a leading paper researcher at North Carolina State University:

> The paper industry is by nature very cautious, but it is aggressively seeking data on hemp, which represents a radical change. . . . By maintaining its hard line against hemp, the legal system throws away both science and common sense.

Plastics

Hemp can be used to manufacture a variety of plastic products. The hurds (short core fibers) may be processed into cellophane packing material, which was common until the 1930s, or they may be manufactured into a low-cost, compostable replacement for styrofoam.

Henry Ford used hemp-and-sisal cellulose plastic to build car doors and fenders. Several German firms are now developing 100-percent-hemp cellulose plastic composites for the manufacture of snowboards and skateboards. An Austrian firm, Zellform, has created a hemp-plastic resin called Hempstone for use in musical instruments, loudspeakers, and furniture. Plant-based plastics made from hemp can be completely biodegradable, and have the potential to reduce oil consumption and the processing of petrochemicals. Hemp-plastic packaging (for use inside cereal boxes, for example) and hemp-foam disposable plates and cups could be composted at home, lowering the high cost of landfilling or recycling petroleum-based plastics.

Chair made of the hemp plastic known as Hempstone — a 100 percent hemp-fiber plastic made by the Austrian firm Zellform.

Textiles

The world's third-largest industry is textile manufacturing. Hemp textiles offer a multiplicity of fabric uses: for bedspreads, blankets, backpacks, carpeting, clothing, draperies, hats, luggage, mattresses, sails, sheets, shoes, shirts, tents, towels, and upholstery, to name only a sampling.

Briefcase made of Hungarian hemp, manufactured in the U.S.

Hemp textiles have a number of distinct advantages over other fabrics. According to the Chinese Academy of Sciences, fabrics containing at least 50 percent hemp block the sun's ultraviolet rays more effectively than do other fabrics. Compared to cotton fibers, hemp fibers are longer, stronger, more lustrous and absorbent, and more mildew-resistant. Hemp fabrics also keep the wearer cooler in the summer and warmer in the winter than do cottons or synthetics.

The hemp-textile industry has been developing yarn that is more lightweight and more uniform than what is currently available. This innovation will allow for the much-anticipated manufacture of hemp T-shirts and other knits. Still another method for making fine textiles from hemp is now being rediscovered in Europe: The "cottonization" process converts long, thick hemp fibers into cotton-like bundles, or hemp flock, which can then be processed through existing spinning and weaving equipment. Companies in the United Kingdom are pioneering the use of this process in blending hemp with fibers such as wool, flax, and cotton to produce fashionable apparel.

OTHER USEFUL FIBER PLANTS

More than fifty plants of other genera are referred to as "hemp," including Manila hemp (*Musa textilis*) and sunn hemp (*Crotalaria juncea*). While these various plants have some useful fiber qualities for twine or matting, they do not yield high-quality textiles, nor do they have the antimicrobial or rot-resistant properties of industrial hemp. Historically, most of these other fiber plants were not used for sails or for marine cordage, because they would mildew and eventually rot when exposed to salt. This is why the United States government has specifically referred to *Cannabis sativa* as "true hemp."

For balanced economic activity, it is important that regional economics rely on a variety of sustainable fibers to supply needed raw materials. For example, according to the Institute for Local Self-Reliance, North America annually burns millions of tons of agricultural straws from corn, rice, and wheat. Following a sixty-year lull, these farm "wastes" are just now starting to be recognized as valuable materials (see chapter 2). Prime Board, a North Dakota company, is transforming leftover wheat stalks into value-added compressed boards for making cabinets and other useful products.

Like hemp, the following plants contain long bast fibers that are suitable for a variety of commercial purposes:

Jute (Corchorus capsularis *and* C. olitorius)

This annual plant ranges from five to sixteen feet tall with stalks from one-half inch to three-quarters of an inch in diameter. Jute is mentioned in the Bible as well as in ancient Egyptian and Mediterranean literature. It grows best in humid, tropical climates, such as India, Pakistan, Bangladesh, China, and Brazil. Historically a strong competitor of hemp for baling twine, jute has other applications including fabrics for bagging and wrapping, as well as nonmarine cordage, carpet backing, linoleum backing, and rugs.

Flax *(*Linum usitatissimum*)*

This plant as well as its fiber are widely known as linen. Flax adapts itself readily to a wide variety of soils and climate conditions. An annual, it grows three to four feet tall with slender stalks up to one-sixth of an inch in diameter. Flax generally is grown in the same plot no more than once every five to six years, partly because it is hard on the soil and is prone to a variety of diseases and fungi. Along with hemp, flax is one of the oldest fibers utilized for cordage and fabrics. Flax is grown for fiber in Belgium, France, Italy, the United Kingdom, the United States, the Netherlands, Hungary, Russia, the Ukraine, and many other former Soviet countries. Different varieties are cultivated for seed production; Canada dedicates the most acreage of any country for this purpose.

Flax and hemp share many similarities. Textiles from flax (linen) and hemp look very much alike, with flax usually having a somewhat finer weave. The seed and oil of both plants have excellent nutritional profiles, including high volumes of essential fatty acids, with hemp oil the richer and longer lasting of the two. Flax is utilized in a wide variety of products comparable to hemp's, including construction materials, paper, and plastic.

Ramie *(*Boehmeria nivea*)*

This multistalked perennial reaches a height of five to eight feet and a diameter varying from one-half inch to three-quarters of an inch. Cloth found on Egyptian mummies dated 5000 to 3300 B.C. has been classified as ramie; written references date back to 600 B.C. in the classical Chinese *Book of Odes*. Ramie grows in moist tropical and subtropical climates, which makes it unsuitable for much of North America. Countries currently producing ramie include India, China, Thailand,

Hemp Seed, Oil, and the Law

For legal importation into countries that prohibit hemp cultivation, hemp seeds must be steam-sterilized for fifteen minutes to make germination impossible—despite the fact that the seed and its oil contain no psychoactive qualities whatsoever. Unfortunately, this practice accelerates rancidity, reduces nutritional qualities, makes the hemp seeds and their derivatives more expensive, and eliminates the option of sprouting the seeds, a key process in creating appealing new hempen foods. Canadians are gearing up for exporting live hemp-seed food and oil products into the U.S. market.

Trace amounts of THC have been found in commercial hemp-seed oil. Although these levels are far below the quantity needed to have any psychoactive effect, some people have found that sensitive urine tests used for employment screening have been affected by the legal use of hemp-seed products. Food manufacturers have discovered that these traces of THC do not come from the seeds, but from the residues of leaves and flower tops that cling to the seeds. Producers of hempen foods now are cleaning their seeds more thoroughly before use, which should alleviate this problem.

South Korea, Japan, the Philippines, Cuba, Haiti, Jamaica, Mexico, Guatemala, El Salvador, Peru, Brazil, and Argentina. There have also been occasional small acreages planted in Florida, the most recently in 1996. Ramie is used for various twine and textile products.

*Kenaf (*Hibiscus cannabinus*)*

Indigenous to Africa, the kenaf plant grows from eight to sixteen feet tall and reaches about one-half inch in diameter. It requires night tem-

La Chanvrière de l'Aube

Freshly crushed French hemp seed oil.

peratures above 50 degrees, yields six to ten tons of fiber per acre, and is ready to harvest in approximately one hundred twenty to one hundred fifty days. Currently farmed in Switzerland, China, India, and the United States, kenaf makes excellent-quality paper, as well as potting mixtures, building materials, and molded products. However, kenaf fiber is not as desirable as hemp or flax for textiles. Because weeds outcompete kenaf, herbicide use is common. Kenaf can be susceptible to diseases and harmful insects.

Urena *(Urena lobata)*

Urena is a herbaceous perennial that reaches ten to fifteen feet, with stalks about as thick as jute. Urena grows in equatorial Africa (particularly Madagascar), India, and Brazil. The fiber has similar purposes to those of jute, and is often blended with it for commercial use.

*Nettle (*Urtica dioica, U. urens, *and* U. pilulifera)

Of these three species of stinging nettle, *U. dioica* is a perennial, and the other two are annuals. Nettle historically has been cultivated in Sweden, Germany, Russia, Italy, and France; product uses have included sails, ropes, and netting. Current world production is very limited due to lack of market demand.

Business Profile

The Rella Good Cheese Company

According to company founder Richard Rose, this venture (formerly Sharon's Finest) began as a food business in 1980 and began to produce hemp foods in 1994. The company today sees annual sales of $3 million.

As the number of natural-foods consumers steadily rises in the general population, The Rella Good Cheese Company is looking to move hemp products into the mainstream. According to Rose:

> We expect to make many breakthroughs within the next year that will revolutionize hemp foods, making it much easier for everyone to use hemp seeds in their diet and recipes. Currently, hemp foods are still in the Dark Ages, but the Renaissance is around the corner.
>
> In the next ten years, I see hemp-food sales increasing exponentially faster than the rest of the hemp industry, and we'll be in the front of the pack. We will consider a public stock offering within five years.

As to why hemp products are in such demand, Rose says:

> People are voting their conscience with their dollars, plus there's a reason it has been used for nine thousand years: It's a very useful fiber. . . . EFAS (essential fatty acids) will become hot, and hemp seed will be the source of choice. People can get this nutrition via the oil or by eating hemp seed. I believe that in five years there will be 150 foods made from hemp seed on the market, in supermarkets as well as natural-food stores.
>
> We introduced HempRella (a hemp-based cheese) at a trade show in Baltimore. We had a booth near the entrance and a big sign. We were swamped for the entire show with people wanting samples and info. After fifteen years in the food business, it was by far the biggest response we had ever had to a new product. That day, the largest distributor told us he absolutely would carry it, and we were off. It was the buzz of the show.

For people thinking about launching a hemp enterprise, Rose offers the following advice:

> The hemp industry will become more professional and competitive. Be ready for that environment or don't get into it. The days of making a few clothes and taking them to Grateful Dead shows are over. Know what you're doing, and continue to strive to do it better.

The Rella Good Cheese Company
P.O. Box 5020
Santa Rosa, CA 95402-5020
Telephone (707) 576-7050
Fax (707) 545-7116
e-mail: richard@rella.com
www.rella.com

Hemp ready for harvesting at Hempline's fields in Ontario, Canada.

I believe that the great Creator has put ores and oil on this earth to give us a breathing spell. As we exhaust them, we must be prepared to fall back on our farms, which is God's true storehouse and can never be exhausted. We can learn to synthesize material for every human need from things that grow.

GEORGE WASHINGTON CARVER

The Farming of Hemp

CHAPTER SIX

THE AMERICAN FARM BUREAU, the Kentucky Hemp Growers Cooperative Association, and the Wisconsin Department of Agriculture are among the legions of respected American farm organizations that support the reintroduction of fiber hemp cultivation in the United States. Why is industrial hemp so attractive to so many farmers?

For centuries, hemp has proven itself to be a dependable low-maintenance crop that is especially useful when planted in rotation with grains, beans, or flax. A fast-growing plant, hemp can reach its full height of six to sixteen feet in as few as ninety days, with little or no need for herbicides or insecticides.

Nevertheless, in spite of hemp's numerous positive commercial features, it presents some significant agricultural challenges. These include an immature market; the lack of a domestic seed supply in many regions, including all of North America; and outdated, inefficient harvesting and processing technologies. Optimally, however, these challenges can be regarded as opportunities waiting for effective solutions, as we will see in this chapter.

From *The Reign of Law: A Tale of the Kentucky Hemp Fields*, James Lane Allen, 1900

Harvest of a turn-of-the-century hemp field.

SUSTAINABLE AGRICULTURE

Hemp is an especially attractive crop to farmers interested in practicing sustainable agriculture, for it thrives on well-manured land, cleans the fields of weeds, returns to soil a high proportion of the nutrients it borrows for growth, and thus leaves the soil in good condition for the crops that will follow.

In 1880, Kentucky scientist Dr. Robert Peter did extensive analyses showing that successive hemp crops could be grown without fertilization when hemp was spread for dew retting (see page 157 for more details on retting) upon the field that had produced it. As the hemp lay exposed to the weather, significant nutrients seeped back into the ground. The land was enriched further by the leaves that fell or were beaten off the hemp stalks, by the roots that remained after the crop was harvested, and finally by the hurds that were separated from the usable fiber. The hurds were considered waste at that time, and were burned in the fields; the ashes were plowed back in for soil restoration. The practice of burning hurds has now been discontinued, due to the value of hurds as short fibers.

No better statement of hemp's beneficial qualities could be made than that of Dr. Lyster Dewey of the United States Department of Agriculture (USDA), who wrote in the agency's 1913 *Yearbook*:

Hemp cultivated for the production of fiber, cut before the seeds are formed, and retted on the land where it has been grown, tends to improve rather than injure the soil. It improves its physical condition, destroys weeds, and does not exhaust its fertility. Kentucky farmers commonly grew hemp in the same fields ten to fifteen years in a row, with the last year being just as productive as the initial ones.

Among fiber crops, hemp is the least chemically dependent, as shown in the accompanying chart. It ranks selected fiber crops with regard to their resistance against pests and weeds.

RELATIVE PEST AND WEED-RESISTANCE OF FIBER CROPS

Crop	Disease and insect problems	Weed problems
hemp	1	1
flax	4	4
kenaf	3	3
cotton	5	5

Relative rank: 0 (better)–5 (worse)   © 1997 HEMPTECH

In April 1995, the U.S. government's National Organic Standards Board drafted a definition that read in part:

> Organic agriculture is an ecological production management system that promotes and enhances biodiversity, biological cycles, and soil biological activity. It is based on minimal use of off-farm inputs and on management practices that restore, maintain, and enhance ecological harmony. "Organic" is a labelling term that denotes products produced under the authority of the Organic Foods Production Act. The principal guidelines for organic production are to use materials and practices that enhance the ecological balance of natural systems and that integrate the parts of the farming system into an ecological whole.

Because of its weed-controlling ability, hemp is an excellent crop to grow on land that is in transition from chemical dependency to certified organic status. Researchers in Poland have demonstrated that fiber hemp can be used as a bioremediation method on land made toxic from industrial heavy-metals pollution. According to the paper "Re-cultivation of Degraded Areas through Cultivation of Hemp" (presented at the 1995 Bioresource Hemp symposium by the Institute of Natural Fibers in Pozan, Poland):

> [H]emp and other fibrous plants tested in the experiment can extract heavy metals from the soil and accumulate them in different parts of the plant. . . . [H]emp takes more metals from the soil than other plants tested.

Perhaps in reaction to the political extremism shown in the complete banning of this valuable crop, there has been a tendency to attribute to hemp the qualities of an *Übercrop*, or Supercrop. This is a forgivable exaggeration: As noted throughout this book and especially in this chapter, hemp does have many unique advantages. But to expect it to save the planet places on hemp an undue burden that we do not place on any other crop. Nor do we legally require that other crops demonstrate their profitability before they can be grown commercially or experimentally.

Hemcore

Processed hemp fiber in Hemcore's warehouse.

MARKETS, MARKETS, MARKETS

In the real-estate business, it is said, the key to success is "Location, lo-cation, location." In the hemp industry, the key is "Markets, markets, markets." To create demand for hemp fiber, market development is es-sential, and several years of consistent effort may be required before buyers will commit to purchasing crops on a regular basis. To ignore this fact can prove to be a costly mistake.

Long before planting any seed, hemp farmers and processors should thoroughly lay the groundwork to find potential buyers for their crops. Otherwise, tons of unshipped stalks may end up as expensive compost.

Remember that hemp seeds, unlike the stalks, can be readily shipped and easily processed into value-added products. Thus, growing hemp primarily for seed rather than fiber is a more conservative approach for farmers venturing into the hemp market, until such time as there are established processing facilities or stable industries ready to pur-chase bulk quantities of hemp fiber (see chapter 5).

THE SEED-STOCK ISSUE

On every continent but Antarctica, there is renewed interest in grow-ing hemp. Unfortunately, the world lacks germ plasm adapted to all of these diverse areas. This certainly is true for the whole of North America, which once boasted several superior varieties that have not been preserved. The famous American hemp, known as "Kentucky Hemp" was primarily of Chinese origin. Plant breeder Dr. David West has concluded that it probably resulted from the mixing of Chinese and European lineages in Kentucky after 1850. USDA hemp breeder Dr. Lyster Dewey bred several improved varieties in the 1920s until his program was terminated by the U.S. federal government in 1933. One of these varieties, *Chinamington*, was the highest-yielding hemp the world had yet seen. It was used as one half of the first hybrid hemp, which was developed by Fleischmann in Hungary. *Chinamington* and the entire Kentucky hemp lineage have been lost, for Kentucky hemp seed stock held by the USDA's National Seed Storage Laboratory in Fort Collins, Colorado, has not been maintained. Representatives of the lab explain that the government's hemp-seed stock was discarded in the 1950s because hemp was no longer considered an important

A Hemp-Seed Repository

The Vavilov Research Institute in Saint Petersburg, Russia, founded in 1902, possesses the world's most complete sampling of *Cannabis* genetic resources. It includes more than five hundred distinct seed varieties from twenty-five countries.

Due to Russian government budget cuts, however, much important seed-preservation work is being postponed. The International Hemp Association (IHA), based in Amsterdam, has been raising funds and providing technical support to the Vavilov Institute in its efforts to preserve this vital collection of hemp germ plasm. Their collaboration is assuring the world that these important *Cannabis* cultivars will not be lost. Individuals and organizations might consider donating funds to the Institute, in care of the IHA (see the appendix for contact information).

fiber resource. Thus, this valuable *Cannabis* germ plasm has been lost to future generations, and American farmers will have to rely on imported seed stock for their initial plantings.

This lack of adapted germ plasm is the chief agricultural impediment to the successful restoration of hemp as a serious industrial crop in the United States. Feral hemp, which grows as a weed in many Midwestern states as a legacy of the "Hemp for Victory" campaign (see chapter 3), is the only genetic legacy of the hemp once widely cultivated here. Americans should be diligently collecting it as a repository of valuable genes; instead, our government seeks it out in order to destroy it. In fact, according to the DEA's own figures, over 98 percent of *Cannabis* plants destroyed in 1993 were not marijauana but hemp.

Tropically adapted *Cannabis* is generally of the relative high-THC variety, and has not been bred as an industrial crop (although breeding to produce tropically adapted industrial varieties would be a goal worth pursuing). Hence the only currently viable option for hemp farmers is to import seed from a very limited selection of breeding programs in France, Hungary, the Netherlands, Ukraine, Russia, and

Feral hemp reseeds annually from World War II-era plantings.

Yugoslavia. Moreover, in the European Union (EU), under regulations that specify a THC threshold of 0.3 percent, the French have maintained a virtual seed monopoly through 1996. The highly productive Hungarian cultivars have been registered with the EU only since late 1996.

When hemp breeders set out to recover adapted germ plasm for North America, we can estimate that it will take them from three to ten years to accomplish their task. Dewey noted that it took three growing seasons for imported Chinese seed to "settle down" in its new Kentucky home. During this settling-down time, growing several successive generations of plants allows natural selection to favor certain ones. Although breeders can assist this process, the development of a new plant cultivar today can take five or more years. In the meantime, North American hemp farmers can expect to use European seed.

Hemp Biology

Industrial fiber varieties can be classified into two basic sexual types: monoecious and dioecious. Monoecious ("mon-EE-shus") varieties have both male and female flowers on each plant; dioecious ("die-EE-shus") varieties have entirely separate male or female plants. The dioe-

Japanese Seed Stock

According to comparison studies done early in the twentieth century by the United States Department of Agriculture, traditional Japanese hemp varieties were quite tall, surpassing Chinese and European ones by reaching heights as great as nineteen feet. Unfortunately, any definitive research that may have been done in Japan on native hemp crops was destroyed, along with most Japanese government records, in the firestorms of World War II.

Since 1946, the Tokyo Metropolitan Government Medicinal Plant Garden has maintained a hemp-seed stock and has bred native varieties for research at a large, secure complex in suburban Tokyo. This facility is surely a valuable cache of information and genetic material, even though its director maintains that the plants are kept solely to teach people what hemp looks like, so they can dispose of it should they find it growing in their area. There have been reports of research and test cultivation of low-THC hemp at Shinshu University in Nagano Prefecture, in Tochigi Prefecture, and in other rural locations.

cious trait is the natural condition for hemp; monoecism occurs infrequently in hemp populations. It was a Russian hemp breeder who first recognized the utility of monoecism in solving two of the plant's critical problems: low seed yield and uneven maturation of the sexes.

The incorporation and stabilization of monoecism in fiber hemp call for the skills of a competent plant breeder. Monoecious hemp varieties require yearly selection to prevent the increasing return of separate male and female plants (that is, dioecism) over successive generations of open-pollinated seed reproduction. Also, because seed from successive reproductions gives progressively lower seed yield in the dual-purpose (fiber- and seed-producing) monoecious crop, it is customary to label monoecious hemp seed according to the number of generations that have been grown since a new supply of seed was obtained from the breeding source.

Hemp's shift into the reproductive phase is regulated by the length of the nights: longer nights favor earlier maturation. The actual timing will vary from one cultivar to another, but it holds true that planting any seed closer toward the equator than its usual cultivation area will accelerate flowering. On the other hand, planting the seed closer toward one of the Earth's poles will delay flowering. This is because during the growing season, in either hemisphere, the hours of darkness diminish as we approach the pole.

Seed Selection

It is risky to take seeds adapted to a certain geographic area and plant them in a different region, because they have not had the opportunity to adjust to local variations in climate, soil chemistry, predation, and so on. For instance, Chilean hemp was grown in North America to extend supplies during World War II, but it performed very poorly in Wisconsin and Iowa—despite the similar latitudes of these states and Chile.

Ultimately, there is no substitute for regional breeding. Growers should bear in mind that until responsible trials have been conducted, performance predictions will require a crystal ball. Moreover, one year's data is inadequate for drawing statistically valid conclusions: Modern plant breeding involves substantial replication of plots across years and locations. In Canada, where experimentation with hemp began in 1994, there still is insufficient data from which to draw good conclusions for cultivar recommendations.

In general practice, the prospective grower should consult a map to determine the latitude where the seed was produced. If this latitude is toward the Earth's pole from where the crop will be grown

(north for Northern Hemisphere growers; south for Southern Hemisphere growers), fiber yields will be low. The grower should therefore consider raising the crop for seed and taking the stalks as a byproduct. On the other hand, if the planting seed originates at a latitude toward the equator, the crop will flower later, so it should be grown for fiber and cut as the male plants begin to shed pollen.

In the first half of the twentieth century, Wisconsin hemp-fiber producers purchased their seed from Kentucky growers because Kentucky hemp had the favorable attribute of delayed flowering in Wisconsin, where summer nights are shorter. Since the stalks could now be harvested before flowering, this Southern cultivar helped Northern farmers solve the problem of uneven maturation rates in male and female hemp plants.

An approach being investigated in Australia, which has a long growing season, is the production of multiple short-season crops from hemp varieties that mature quickly in Australian latitudes. This method makes it possible to harvest large total annual yields of fiber for paper, construction materials, or other end uses.

When purchasing hemp seed, the grower should inquire as to the following:

- the zone of adaptation;
- whether the variety is dioecious or monoecious; and
- the reproduction number, for monoecious varieties.

Monoecious planting seed should be "elite"—direct from the breeding source—or "first reproduction"; second reproduction or later should be avoided, if possible, since seed yields will be lower with successive generations.

Once the seed has been selected, the next step is to increase the seed supply for the needed acreage. A hemp-seed field planted at a rate of 10 pounds to an acre, on approximately 20-inch spacing, should yield 700 to 1,000 pounds of seed per acre. The monoecious and unisexual hybrids produce the most seeds.

Dr. Ivan Bócsa, the eminent Hungarian hemp-seed breeder, has experimented with many hybrid varieties in search of higher yields. Bócsa coined the term "unisexual" to mean the generation of seed resulting from the cross of females of a dioecious population with pollen from a monoecious pollinator. His *Uniko-B* is a cross of *Fibrimon* by *Kompolti*, the dioecious Hungarian hemp variety of Italian lineage. *Uniko-B* and *Kompolti* both have performed well in trials in Ontario, Canada. However, legal complications involving THC content have limited the range of seed cultivars permitted in Canadian trials. Another Hungarian hybrid, *Kompolti Hybrid TC*, is a three-way cross of Chinese lineage. The parent plant is notable for its exceptional (for hemp) seed yields: as much as fifteen hundred pounds per acre. So far, however, slightly higher THC content has impeded the use of this variety.

Uncultivated hemp of ancient China had an average fiber content of around 12 to 15 percent. Today's hemp varieties have twice as much long bast fiber. However, high-fiber varieties can slip to lower fiber percentages over successive generations of reproduction. For these reasons—to maintain high fiber percentage, high seed yield, and ul-tra-low THC content—serious hemp growers return directly to the breeding source on an annual or biannual basis for elite seed (seed of highest performance potential). Large-scale hemp farming will re-quire the constant participation of plant breeders and the commercial seed industry to ensure viable seed reproduction and selection.

For most other crops, farmers have annually repurchased seed from firms that bear the high research costs of cultivar improvement. There is no reason to expect this system to be any different for hemp, once the plant's legality has been reestablished. Accordingly, advocates of open-pollinated nonhybrid seeds, or "heirloom seeds," are concerned about the established seed-industry practice of selecting varieties pri-marily for yield—without consideration for flavor, nutritional quality, or disease resistance. How this issue plays out will probably be deter-mined by which companies gain control of the revitalized hemp-seed industry.

Robert C. Clarke/IHA

Valuable hemp seeds ready for planting.

RAISING A HEMP CROP

Hemp is a relatively easy crop to grow, and farmers need follow only a few established guidelines to be rewarded with high yields. Additionally, hemp farmers receive the benefit of improved soil condition for the next crop planted.

This chapter provides a basic overview on farming hemp, yet it is not intended to be a detailed treatise. Potential hemp growers may want to read the authoritative book *Cultivation of Hemp: Botany, Varieties, Cultivation, and Harvesting* by Dr. Ivan Bócsa and Michael Karus (to be published in English by HEMPTECH in 1998).

Crop Rotation

Like corn, hemp is an annual plant, and must be seeded every year. Traditionally, hemp has been grown in rotation with corn, a small grain such as wheat, and a multiseason legume such as alfalfa or clover. English farmers are reporting 10 percent increases in yield when winter wheat is planted following hemp.

USDA studies conducted at Iowa State University in 1943 and published in the 1944 USDA bulletin *Hemp Production Experiments* concluded: "The relative yield of hemp following different crops was found to be in the same order as the effect of these crops on the supply of available nitrogen in the soil." In other words, hemp grown after alfalfa has a higher yield than hemp grown after soybeans, and the yield of hemp grown after soybeans is higher than that of hemp grown following corn or small grains. Because hemp fits nicely into standard crop rotation plans used by cornbelt farmers, it makes an excellent addition to these growers' repertoire of crops for a diversified agriculture.

Soil Selection

Any land that grows good corn will produce good hemp, which thrives in well-drained soil with proper tilth (texture) and ample organic matter. Traditionally, hemp has flourished when planted on the best land, such as the deep black grassland soils of the Ukraine-Russian plateau, which are kept highly fertile through copious manuring.

Hemp generally does not do well in light soils, marginal soils low in organic matter, or soils that are poorly drained.

Moisture Requirements

Commercial growers do not consider hemp a drought-tolerant crop. Areas where it has typically done well receive annual precipitation of about thirty inches. Water is most important to the crop during the first thirty to forty days of growth, when the plants can grow twelve inches in a week. But hemp does not like to stand in water. Dr. Lyster Dewey wrote:

> Very few farm crops require so much water as hemp, yet it will not endure standing water about its roots. In soils of good capillarity, where the general level of soil water is within 10 feet of the surface, there is little danger of injury from drought after the first thirty days, during which the root system of the hemp plant will become well established.

The American Midwest, where the crop historically performed well, typically receives three to four inches of rain during each month of the growing season. Dry conditions later in the growing season are less serious. In fact, the USDA reported on the positive impact of the 1930 drought on that year's hemp crop:

> This year's drought has had a good effect on one crop at least. Hemp grown in Kentucky, Illinois, and Wisconsin, in fields that were well prepared and planted early, is giving a remarkable yield of fiber of exceptionally good quality, say the fiber plant specialists of the Department. Some of Wisconsin's 1930 crops show a yield of about one thousand pounds of long fiber an acre; the average yield in that state for the preceding ten years was about eight hundred pounds an acre.

Under such circumstances, hemp's long taproot is able to draw from subsoil moisture. In drier climates, irrigation may be necessary (presuming that the moisture is there to be drawn from).

Hemp grown in California early in this century was irrigated, and yielded twice as much fiber (one whole ton compared to one-half ton per acre) as did Wisconsin crops of the same period. The ill-fated 1994 Imperial Valley crop of the Hempstead Company, plowed under by overly zealous California government officials just before harvesting, also was irrigated.

Obviously, in most areas, irrigation will add to a hemp farmer's production costs. The question of how irrigated and unirrigated hemp

Robert C. Clarke/IHA

Hemp's deep taproot aerates soil and adds valuable organic matter.

fields will compete economically in a modern setting may require additional research.

Temperature Requirements

Hemp does relatively well in the face of temperature extremes. When a hemp seed germinates, it gives off heat, metabolizing its stored oil. A 1914 experiment published in the *Botanical Gazette* by Darsie et al. found hemp to be the hottest shoot of those tested.

It is generally advised that growers plant hemp one to two weeks before it is time to plant corn. Hemp will withstand light frosts, and it grows at colder temperatures than kenaf.

Geof G. Kime

Cross-section of a thin hemp stalk.

Seedbed Preparation

Every discussion of the agronomics of hemp stresses the critical importance of good seedbed preparation, because this supports the uniform growth of the stalks. Uniformity is a key factor for good fiber: Overly large stalks will yield a coarse, low-grade fiber, whereas stalks one-quarter inch in diameter will produce a high-grade fiber. Thorough harrowing of the seedbed—to break up large clods and smooth the soil surface—is recommended in order to facilitate uniform stalk development.

Seed-planting methods also affect uniformity. In the old days, hemp was sown by broadcasting, that is, throwing the seeds onto the soil. This tactic produced an uneven stand, since the seeds were spaced randomly. The preferred modern method is to use seed-drilling equipment, which spaces the seeds in a designated regular pattern and results in much more uniform stalk diameters.

A crop of hemp is actually a dense population of genotypes, or genetically diverse individuals. Pressure from the dense plant growth—the same quality that makes hemp a good weed eliminator—forces uniformity on individual stem development. Where plant density is compromised by spatial gaps, uneven fertility, or other disturbances,

Drilling hemp seeds into prepared seed bed.

opportunistic genotypes will grow excessively and uniformity will decrease.

What about hemp's role in no-till agriculture? Because this farming technique is a recent development, hemp and no-till have not met. No-till cropping is generally chemical-intensive for weed control, and in fact the technique is made possible by the liberal use of herbicides. But it is conceivable that hemp could be integrated into a no-till system in a way that could reduce the herbicide requirement. It will be exciting to see what stewardship-oriented farmers will figure out once they can freely experiment with industrial hemp.

Although some growers have tried planting hemp into alfalfa stubble, traditionalists do not recommend it. The more conservative ap-

"... before the red of apple buds become a sign in the low orchards, or the high song of the thrush is pouring forth far away at wet pale-green sunsets, the sower, the earliest sower of the hemp, goes forth into the fields."

from The Reign of Law: A Tale of the Kentucky Hemp Fields,
James Allen, 1900.

proach has emphasized the critical role of smooth fields of well-tilled soil, again for the purpose of engendering uniform stalks and high-quality fiber.

Fertilizer

Hemp is appealing in that it lends itself to organic farming methods, but if and when the crop achieves the planting acreage required to supply paper or fiberboard factories, it seems unlikely that most growers will be organically inclined. Commercial chemical fertilizers can be applied on hemp as on corn; hemp's requirement of nitrogen, phosphorus, and potassium is similar to that of corn when the two plants are grown under similar conditions.

Experienced hemp farmers know that pouring nitrogen on a crop will not result in ever-higher yield or quality. It is better to err on the light side than on the heavy side with inorganic nitrogen. Overfertilization is associated with excessive growth, weaker fiber, increased self-thinning, and less uniform stalks. During World War II's emergency hemp production, chemical nitrogen was in short supply. Yet hemp grew fine without synthetic fertilizers where the land was rich in organic matter, either from spreading barnyard manure or planting after a crop of legumes. Since hemp does not fix nitrogen, legumes such as alfalfa and clover should be planted in rotation to do so, and to keep the soil fertile. In fact, hemp takes more nutrients from the soil than it returns, although with field-retting the balance sheet becomes attractive: Hemp returns in its leaf litter approximately half of the nitrogen it consumes for growth. Nevertheless, this means that the crop removes a corresponding amount of nutrients, which disproves the frequent assertion that hemp requires no fertilization nor other soil amendments.

One hundred years ago, after visiting France, Charles Dodge of the USDA's Office of Fiber Investigations reported:

> A rotation of crops is practiced, hemp alternating with grain crops, although . . . it is also allowed to grow continually upon the same land. Regarding this mode of cultivation, they [the French] consider it is not contrary to the law of rotation, as by deep plowing and the annual use of an abundance of fertilizers the ground is kept sufficiently enriched for the demands that are made upon it. If the soil is not sufficiently rich in phosphates or the salts of potassium, these must be supplied by the use of lime, marl [calcium carbonate deposits], ground bone, animal charcoal, or ashes mixed with prepared animal compost. Even hemp-cake, the leaves of the plant, and the "shive" or "boon" [short core fibers] may be returned to the land with benefit. This high fertilizing is necessary, as the hemp absorbs the equivalent of 1,500 kilos of fertilizers per every hundred kilos of fiber obtained.

USDA studies at Iowa State University in 1943 found no effect for added potassium to either low-or high-yielding soils, but "soils well supplied with nitrogen and organic matter gave the highest yield of hemp."

Seeding

Generally speaking, soil temperature should reach 8 to 10 degrees Celsius (45 degrees Fahrenheit) before hemp is sown.

Young hemp plants often grow one inch or more each day.

Hemp plants a few weeks after emerging from the ground.

Thin stalks, which grow up to 400 plants per square meter.

Seeding rates for hemp vary widely. In the Ferrara region of Italy, hemp traditionally was sown at a rate of hundreds of pounds of seed per acre. This approach produced a flax-like hemp famed for its fine quality. In the United States, the rate recommended by the Wisconsin hemp industry during World War II was a bushel and a peck (fifty-five pounds) per acre. However, rates as low as thirty-five pounds of seed per acre reportedly have given yields approximating those produced by higher rates.

Imported hemp seed is expensive, so the lower sowing rate is recommended if the goal is an industrial fiber crop. Finer, textile-quality hemp requires higher seeding density.

The best results have been obtained by planting hemp with a grain drill or Brillion seeder with four-inch row spacing. Farmers should test seed germination and adjust the planting rate accordingly.

Geof G. Kime

*Hemp's dense planting smothers weeds, so the field
is left virtually weed-free for the following crop.*

Weed Control

An appealing characteristic of the hemp crop is its ability to discourage the presence of weeds, due to the shade created by its dense growth. Andrew Wright, an early-twentieth-century agronomist with the University of Wisconsin Agricultural Experiment Station, attributed the enthusiasm of Wisconsin farmers for hemp to this side effect of its production. Testimonials to hemp's weed-controlling capability can be found in every treatise on hemp agriculture, including Wright's 1918 article "Wisconsin's Hemp Industry":

> Hemp has been demonstrated to be the best smother crop for assisting in the eradication of quack grass and Canada thistles. . . . At Waupon in 1911 the hemp was grown on land badly infested with quack grass, and in spite of an unfavorable season a yield of two thousand one hundred pounds of fiber to the acre was obtained and the quack grass was practically destroyed.

However, hemp's ability to suppress weeds does not mean that it can be planted without proper preparation of the land. Weedy land should be thoroughly plowed and disced to destroy weeds prior to planting. Given a head start, hemp will progressively reduce weed pressure for following crops. To bring persistent weeds under control, one should initially plant hemp for two years in succession on the same land. After that, normal rotation of crops will keep weeds down.

This characteristic of hemp was demonstrated in the Netherlands by Lotz, Groeneveld, Habekotté, and van Oene, whose 1991 study of hemp's suppression of yellow nutgrass (*Cyperus esculentus*) concluded:

> [H]emp was the most competitive crop in this study. Selecting this crop in a rotation will cause the strongest population reduction of *C. esculentus* on infested farmland. This control option of hemp against such harmful weeds as *C. esculentus* is an attendant benefit of the introduction of hemp as a commercial crop.

In the first year of the nutgrass experiment [1987], fields infested with this weed were planted to one of four crops—maize [corn], hemp, winter barley, or winter rye—and one field was left fallow. In the following two years, all fields were planted with maize. The incidence of the weed was counted before and after each rotation, and was significantly reduced only when the preceding crop was hemp.

Diseases and Infestations

In June and December 1996, the *Journal of the International Hemp Association* published a two-part review of *Cannabis* diseases and pests by John McPartland, who wrote:

> *Cannabis* has a reputation for being pest-free. Actually, it is pest-tolerant. . . . Most *Cannabis* pests are insects. Nearly three hundred insect pests have been described on hemp and marijuana, but very few cause economic crop losses. In hemp crops, the most serious pests are lepidopterous stem borers, predominately European corn borers (*Ostrinia nubilalis*) and hemp borers (*Grapholita delineana*). Beetle grubs also bore into stems and roots (e.g., *Psylliodes attenuata*, *Ceutorhynchus rapae*, *Rhinocus pericarpius*, *Thyestes gebleri*, and several *Mordellistena* species).
>
> The claim that *Cannabis* has no diseases is not correct. *Cannabis* suffers over one hundred diseases, but less than a dozen are serious.

COMMON CANNABIS PESTS

Seed and seedling	Flower and leaf, outdoors	Flower and leaf, indoors	Stalk and stem	Root
cutworms	hemp flea beetles	spider mites	European corn borers	hemp flea beetles
birds	hemp borers	aphids		white root grubs
hemp flea beetles	budworms	whiteflies	hemp borers	root maggots
crickets	leaf miners	thrips	weevils	termites and ants
slugs	green stink bugs	leafhoppers	modellid grubs	fungus gnats
rodents			longhorn grubs	wireworms

SOURCE: *Journal of the International Hemp Association* 3, no. 2 (1996).

It has yet to be documented that chemically treating an insect or disease problem in hemp is economically justified. Investigations in the Netherlands in the early 1990s concluded that chemical treatment for the leaf fungus *Botrytis*, known to vintners, was not necessary. (Besides, there are organic methods for controlling *Botrytis*.) Other Dutch research, published by Kok and Coenen in 1994, found that hemp is a poor host for the destructive nematode *Meloidogyne chitwoodi*, meaning that hemp grown on infested soil actually retards the proliferation of this particular pest. Since *M. chitwoodi* is a severe threat to potatoes, incorporating hemp into a rotation with potatoes can help to reduce damage to that crop.

This is yet another example of how the inclusion of hemp can enhance the value of other crops produced on the farm, adding to farm income in indirect ways. Nevertheless, hemp is not invincible, as McPartland has pointed out. When seed brought from another environment has not been bred locally, there is always a chance that the hemp crop will be seriously susceptible to unfamiliar pathogens. Also, when hemp is planted at a late date or on marginal land, a host of factors can contribute to a failed crop. This was the experience in Manitoba in 1995, as reported by Dr. Jack Moes of the Crop Development Section, Manitoba Department of Agriculture:

> By the time the necessary permits were in hand and the seed was imported, it was early June—three to four weeks after the ideal seed-

COMMON CANNABIS DISEASES

Seedling diseases	Flower & leaf diseases, outdoors	Flower & leaf diseases, indoors	Stalk and stem diseases	Root diseases
damping-off fungi	gray mold	nutritional diseases	gray mold	fusarium root rot
storage fungi	yellow & brown leaf spots	pink rot	hemp canker	root knot nema
genetic sterility	downy mildew	gray mold	fusarium canker	broomrape
	olive leaf spot	powdery mildew	fusarium wilt	rhizoc root rot
	nutritional diseases	brown blight	stem nema	sclerotium rot
	brown blight	virus diseases	charcoal rot	cyst nema
	bacterial leaf diseases		anthracnose	
			striatura ulcerosa	
			dodder	

Source: *Journal of the International Hemp Association* 3, no. 1 (1996).

ing date. Bertha armyworm (*Mamestra configurata*) loves hemp leaves and is capable of reducing a plant to nothing but a stalk with a few leaf skeletons—some salvageable fiber, but no further yield. Bertha is a periodic pest in Manitoba, and after a year or two is not likely to be much trouble for another ten to fifteen years.

Some suggest that hemp is free of diseases. However, it is quite susceptible to Sclerotinia (*Sclerotinia sclerotiorum*), which came in the form of stem lesions that killed off the upper part of the plant and left the plant very prone to lodging [falling over].

As a result of the late planting, the Bertha infestation, and the sclerotinia, several of the department's test plots were lost that year.

Comparatively speaking, however, hemp certainly is one of the most pest-tolerant crop plants. The Chinese surround rows of vegetable crops with hemp to repel whiteflies and other pests; on the other hand, the European corn borer has been known to burrow hemp stem. When planting hemp in rotation with corn, growers should take this possibility seriously and watch carefully for early detection of any infestation. In Wisconsin, the European corn borer rarely overwinters, so no build-up problem would be anticipated, but in Southern areas of the United States, this insect could pose a problem.

Geof G. Kime

Hemp stalks retting in the field in Ontario, Canada.

Geof G. Kime

Turning hemp stalks to insure more uniform retting.

HempFlax

A HempFlax hemp harvester's circular drums cut and chop hemp stalks.

Harvesting

Though farmers love hemp's strength and the speed with which it grows, they know that its tough fibers make it a harvesting challenge. Ian Low of the United Kingdom's Hemcore likes to joke about how hemp stalks have broken nearly every type of harvesting equipment that his suppliers have used. Much of the existing hemp-harvesting technology is just not up to par. The harvesting is still done by hand in most of Asia, with 1940s-era equipment in Eastern Europe, and with modified farm implements in Australia, Western Europe, and North America. Although International Harvester developed hemp-harvesting equipment in the 1930s, it has been close to sixty-five years since a major farm-equipment manufacturer addressed this issue.

It is good news, therefore, that a Dutch company, HempFlax, has recently developed several new kinds of specialized harvesting equipment, including a self-driven square bale press and a wide-mowing hemp harvester with the capacity to process 4.5 to 6.5 acres per hour. A majority of today's hemp farmers still use a modified sickle cutter; later, after the stalks are retted, the farmers use a square or round bale harvester to gather the fiber for shipment to a hemp mill.

Geof G. Kime

A German baler gathers retted hemp into square bales ready for market.

A modified combine harvesting a hemp fiber and seed crop in Austria.

Since hemp-harvesting methodology depends on the end use intended for the crop, the details of harvest techniques will be specified by the end user: a paper mill, a regional processor, or a growers' co-op such as the Kentucky Hemp Growers Cooperative Association, for example.

Hemp seed may be taken with a combine that also processes the fiber. Modified or specialized seed-harvesting equipment is needed, since hemp grows up to sixteen feet tall; no other major annual crop attains such heights.

Retting

Retting is the traditional process of letting the hemp stalk partially rot in order to separate the bast fiber from the core. It is a critical step in the handling of the hemp crop. Retting has been done in many ways: in streams and ponds, in tanks and artificial pools, on land, and in snow. For textile-grade fiber, the process has historically involved water retting, which has environmental disadvantages. Water retting in

Instit. of Field and Vegetable Crops, Novi Sad

Yugoslavian worker assessing fiber quality in a traditional water retting hemp pond.

Geof G. Kime

Author holding retted hemp stalks in Ontario, Canada.

tanks tends to be anaerobic, and creates an offensive odor. The water must be disposed of in an environmentally sensitive manner, for it can be a nasty pollutant.

The preferred method in North America was field dew retting, in which the cut stalks are left out in the field for several weeks while natural humidity and bacteria decompose the fiber-binding pectins. This method is dependent on adequate fall moisture; an entire year's crop was lost in Wisconsin in the 1950s due to poor weather conditions. On the other hand, field retting does return nutrients to the soil as the hemp decomposes.

Regardless of retting technique, the process must be monitored carefully so that it progresses far enough without going too far. When the adequately retted stalk is flexed rapidly, its core should readily separate from the long fibers.

There have been several experiments with the decortication of unretted fiber, as in 1917, when George Schlichten ran a large hemp-processing business in California that used only unretted (green) hemp. For modern industrial applications such as medium-density fiberboard, it is possible for the hemp stalk to be used whole. The crop can be harvested with modified forage balers or chopped into short lengths and transported to the manufacturing site, eliminating the retting requirement.

In order for industrial hemp to realize its full economic potential, advances will have to be made in several stages of processing hemp fibers. New methods are being explored, including ultrasonics and steam explosion (first tried in the 1920s in Germany), that may make the retting process obsolete. In the coming decades, it will be exciting to watch the development of new methods and technologies to grow and process this ancient crop.

Geof G. Kime

Hempline founders Joe Strobel and Geof Kime in Ontario, Canada hemp field.

Hempline

Hempline, Inc., was formed out of a desire to develop hemp as a profitable, renewable natural resource for Ontario. The founding directors, Joe Strobel and Geof Kime, were interested in finding an alternative crop for tobacco. At the time, it was illegal to grow hemp without a license.

In 1994, the year of Hempline's incorporation, the company applied for its first license to grow ten acres of hemp in the tobacco-growing region near Tillsonburg, and subsequently harvested the first legal crop of North American hemp in modern times. Hempline has since developed its own hemp-harvesting and fiber-separation technologies.

> We are expanding the capacity of our fiber-separation process, and will be increasing the acreage of hemp grown close to the mill. Now that hemp fiber is no longer regulated in Canada, we will be developing new markets for hemp textiles, specialty paper, composites, and fiberboard.

Kime explains why Hempline now supplies hemp fiber to numerous American firms:

> There are many reasons why hemp products are of such interest again, including the strength and absorbency of the fiber, the perceived environmental and economic benefits, and a mystique and allure from the long prohibition on hemp's cultivation.

Hempline, Inc.
10 Gibson Drive
Tillsonburg, Ontario N4G 5G5
Canada
Telephone: (519) 434-3684
Fax: (519) 434-6663
e-mail: fiber@hempline.com
www.hempline.com

Not only is hemp fast-growing — it sequesters carbon dioxide from the atmosphere and incorporates it into useful fiber products. In this way hemp plays an important role in mitigating the greenhouse effect.

*The twentieth century has been the age of the hydro-
carbon. The twenty-first century should witness a
rebirth of a carbohydrate economy. Living plants are
again becoming attractive raw materials for manufacturers. The signs may be modest, but the conclusion is
unmistakable. The pendulum is swinging to
a biological economy.*

DAVID MORRIS AND IRSHAD AHMED
Institute for Local Self-Reliance

Hemp and a Sustainable Future

CHAPTER SEVEN

WHEN ONLY A FEW hundred million people inhabited the Earth, the conservation of natural resources for future generations was not of global concern. Now, however, be- cause our current population is close to six billion (and expected to reach ten billion in the twenty-first century), the question of how to sustain our dwindling natural resources is becoming critical for all of humanity.

"Sustainability" has different meanings for different people. Its root verbal meaning is "to support; to nourish; to keep going, as an ac- tion or process." Since the 1980s, scientists and proponents of resource conservation have made sustainability a benchmark for planetary well-being. This concept enables natural systems to continue func- tioning well into the future.

RECYCLING IN NATURE

Imagine we are walking through a forest of Douglas firs, cedars, madrones, and tan oaks, all of various ages. Huge fallen trees rest on the forest floor, slowly decomposing into rich, crumbly soil. We see

Consuming the Capital

Clearly, any human activity dependent on the
consumptive use of ecological resources (forestry,
fisheries, agriculture, waste disposal, urban
sprawl on agricultural land) cannot be sustained
indefinitely if it uses not only the annual production of the biosphere (the "interest") but also cuts
into the standing stock (the "capital"). Herein lies
the essence of our environmental crisis.
                                        —William E. Rees

that during the winter rains, a new seedling has sprouted from what
was formerly a strong, standing fir. Three hundred years from now,
that seedling may well be as large as the huge fallen fir that served as
its seedbed, and may one day provide nutrients, in turn, for the next
generation of trees.

The Earth's few remaining old-growth forests serve as prime examples of sustainability. When not interfered with, these naturally
evolving forest systems have become completely self-sustaining over
time. Complex interactions in microscopic soil life, and unique lichens
that grow on branches two hundred feet above the forest floor—all are
part of the miraculous web of life.

Like the soil organisms, humanity, too, can play a positive role in
nature. Our recycling efforts today are an awkward attempt to sustain
the human race and the mineral, plant, and animal kingdoms, as we
save and reuse materials with an eye to the future. By observing nature, the master recycler, we may learn to incorporate her valuable
lessons into our daily lives so we can conserve resources for our children's children.

Yet, as Paul Hawken, author of *The Ecology of Commerce* and the
forthcoming *Natural Capitalism: The Coming Efficiency Revolution*
(with Amory and Hunter Lovins), pointed out in *Mother Jones,* April
1997, our industrialized society discounts the value of natural systems:

You can win a Nobel Prize in economics and travel to the Royal Palace
in Stockholm in a gilded, horse-drawn brougham believing that
ancient forests are more valuable in liquidation—as fruit crates and

Yellow Pages—than as a growing and growing concern. But soon (I would estimate within a few decades), we will realize collectively what each of us already knows individually: It's cheaper to take care of something—a rope, a car, a planet—than to let it decay and try to fix it later.

While there may be no "right" way to value a forest or a river, there is a wrong way, which is to give it no value at all. How do we decide the value of a 700-year-old tree? We need only ask how much it would cost to make a new one. Or a new river, or even a new atmosphere.

When we fail to follow natural cycles, which really are fundamental life-operating principles, unintended and sometimes drastic results can occur. For every action there is a reaction. For example, the conversion of the Central and South American coffee industry from primarily small, forestry-grown operations to large field-grown plantations has markedly reduced the shade-canopy habitat in these areas. As a result, North Americans have been seeing fewer and fewer orioles and other songbirds that winter in the diminishing coffee-forest habitat.

HOW NONSUSTAINABILITY AFFECTS PEOPLE

Critics of the environmental movement sometimes say that its proponents care only about nature—not about people. Yet, when a region's natural resources are depleted, the end result for many of its human inhabitants is poverty, suffering, displacement, or even death. In "Strategies for a Sustainable World," an article in the January–February 1997 *Harvard Business Review*, Stuart L. Hart writes that in China,

> an estimated 120 million people now roam from city to city, landless and jobless, driven from their villages by deforestation, soil erosion, floods, or droughts. Worldwide, the number of such environmental refugees from the survival economy may be as high as five hundred million people.

One doesn't need a Ph.D. in biology to figure out that we are facing untold dangers because of the way we live in these times. Here are but a few examples of ecological stresses upon the planet:

- Huge underground aquifers—the only source of fresh water in some regions—are being depleted in China, India, the Mideast, the United States, and elsewhere.

- Forests worldwide are being cut down at a faster rate than they can regrow, including ancient forests whose gifts cannot be recovered for centuries.
- Frogs, toads, and other amphibians—bellwethers of air and water quality—are disappearing from their habitats all over the globe.
- Approximately six pounds of topsoil washes away into rivers and oceans for every pound of food produced in the United States.
- Certain toxic pesticides are banned in the United States, yet they are sold to other countries (by American companies, among others) and return to all who eat them in foods sprayed with precisely these banned chemicals.

In the face of all this environmental degradation, hope can be found in the notion of sustainability—hope for more than just human and planetary survival. Stuart Hart goes on to say in his article: "Those who believe that ecological disaster will somehow be averted must also appreciate the commercial implications of such a belief: over the next decade or so, sustainable development will constitute one of the biggest opportunities in the history of commerce."

Consumption for Its Own Sake

Bill Bennett, a prominent conservative, lamented in a late 1996 speech:

> While America is objectively the most powerful, affluent, and envied nation in the world, and in many ways justifiably so, America also leads the industrialized world in rates of murder, violent crime, juvenile violent crime, imprisonment, divorce, abortion, single-parent households, teen suicide, cocaine consumption, per capita consumption of all drugs, pornography production, and pornography consumption.

An expanding body of knowledge confirms these sad facts, as well as the fact that the consumption of resources by the developed world cannot continue at the present rate. Marketeers have sold the American dream beyond their own wildest imaginings, and we are obsessed with new fads, styles, and symbols of success. With television reach-

The McDonough Principles

One of the world's most innovative designers in resource productivity is William McDonough, dean of the University of Virginia's School of Architecture. Inspired by the way living systems actually work, McDonough follows three simple principles when redesigning processes and products:

1. *Waste equals food.* This principle encourages the elimination of the concept of "waste" in industrial design. We need to design every process so that the products themselves—as well as leftover materials and effluents—can become "food" for other processes.
2. *Rely on current solar income.* This principle has two benefits. First, it diminishes, and may eventually eliminate, our reliance on hydro-carbon fuels. Second, it means designing systems that sip energy instead of gulping it down.
3. *Respect diversity.* We need to evaluate every design for its impact on plant, animal, and human life. For a building, this means, literally, what will the birds think of it? For a product, it means, where will it go and what will it do when it gets there? For a system, it means weighing immediate and long-term effects and deciding whether these enhance people's identity, independence, and integrity.

ing every country in the world, billions of people everywhere believe that material possessions equal satisfaction, and are eager to follow in American footsteps. Families from Bogotá to Beijing, watching seductive American programming, don't comprehend the dark side of our consumerism.

Where Have Your Blue Jeans Been?

The following example illustrates how one individual's choice regarding something as simple as buying a new pair of blue jeans can make a difference in global sustainability. Let's compare the effects of purchasing a pair of regionally manufactured jeans instead of a pair made in a distant part of the world.

In the first instance, a farmer grows some hemp and sells it to a nearby regional processor who offers well-paying jobs. The processor markets the resulting hemp textiles to a local garment factory, where a pair of hemp jeans is made. These jeans are then resold at a family-owned retail store in the same area. The net effect of this sequence of events—the growing, processing, manufacturing, and retailing, all in one region—helps the local economy by creating jobs and circulating money within the community. The buyer of these hemp jeans thus contributes to the well-being of his or her region, without needlessly consuming fuel resources to transport goods over long distances. (Of course, some areas may not have enough planted acreage to support a textile-spinning factory. In these cases, the long hemp fibers will have to be shipped to the nearest regional processor.)

Meanwhile, an industrial cotton farmer is under pressure from an international textile company to reduce prices. The grower applies more chemicals to try to boost yields, thereby polluting the groundwater. The textile company manufactures a pair of jeans using cheap, hazardous dyes and maintaining sweatshop conditions for its workers, and ships the jeans thousands of miles across the sea. In the country of import, further transportation is required from the port of entry.

The quantity of petroleum involved in moving consumer goods around the globe is phenomenal. To meet this demand, major oil companies are drilling in the oceans and rain forests, where repeated oil spills kill fish and other wildlife. Hundreds of native cultures worldwide have disappeared due to the blatant violation of their land rights in the rush for oil, fiber, and minerals. Corporate-backed governments do not recognize that their indigenous peoples, who have lived in sacred balance with the waters and forests for eons, should have input on how land is managed.

Nevertheless, while international trade will always play an important role in our lives, individual communities can positively influence local economies. Areas where factories will be shut down due to the

An Early Admonition

From the foreword by Henry A. Wallace, U.S.
Secretary of Agriculture, to the 1938 *Yearbook
of the United States Department of Agriculture*:

> The earth is the mother of us all—plants, ani-
> mals, and men. The phosphorus and calcium of
> the earth build our skeletons and nervous sys-
> tems. Everything else our bodies need except air
> and sun comes from the earth.
>
> Nature treats the earth kindly. Man treats her
> harshly. He over plows the cropland, overgrazes
> the pastureland, and overcuts the timberland. He
> destroys millions of acres completely. . . . The
> flood problem, insofar as it is man-made, is
> chiefly the result of over plowing, overgrazing,
> and overcutting of timber.
>
> This terribly destructive process is excusable
> in a young civilization. It is not excusable in the
> United States in the year 1938.
>
> The social lesson of soil waste is that no man
> has the right to destroy soil even if he does own
> it in fee simple. The soil requires a duty of man
> which we have been slow to recognize.

impact of global trade agreements can effectively weatherize their economic houses against financial storms by pursuing bioregional development strategies. For example, Germany is currently preparing to grow, process, manufacture, and market hemp jeans in local regions. A growing number of informed farmers, workers, and businesspeople around the world are developing a strong local hemp industry that will benefit not only their own and their children's lives, but the lives of many others in their communities as well.

RETHINKING OUR MAJOR RAW MATERIALS

As the number of conscious and informed producers and consumers increases within the marketplace, there will be ever more demand for goods and services provided in a more sustainable manner: products

Total Pesticide Applied Per Cotton Acre in California, 1990–1994

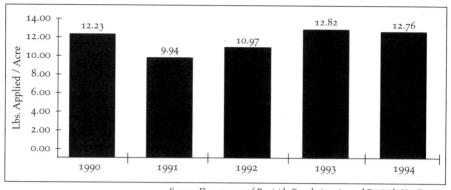

Source: Department of Pesticide Regulation, Annual Pesticide Use Report.
Note: 1990 was the first year for 100% pesticide use reporting in California.
See page 206 for a listing of major cotton pesticides used in the U.S.

made from organic cotton, recycled fibers, industrial hemp, agricultural straws (including those from rice, wheat, barley, and corn), and sustainably harvested wood products. This trend is essential as well as promising, because the rate at which and manner in which we consume cotton, timber, and petroleum is nonsustainable to the point of acting as a destabilizing force.

Cotton

Cotton fields are sprayed with vast quantities of herbicides, insecticides, fungicides, and defoliants. Although total cotton acreage constitutes less than 1 percent of the U.S.'s agricultural lands, it consumes approximately 25 percent of the pesticides—the most of any production crop. Many of these synthetic compounds are extremely poisonous to humans, in even minute quantities: Farmworkers routinely encounter drifting chemicals from crop dusters, besides the daily exposure from working and living next to the fields. Such long-term contact with these substances can lead to cancer, birth defects, and other health problems. In 1994, a pesticide commonly used for cotton, metam sodium, spilled into California's Sacramento River, where it killed nearly everything over a twenty-mile stretch. Much of the groundwater tested in agricultural regions around the world has shown contamination by farm chemicals.

A 1996 report by Agricola Partners of Davis, California, entitled "Pesticide Use in Cotton Growing," states:

These chemicals account for more than 50 percent of the total cost of cotton production in much of the world. . . .
    In 1992, a total of 71.5 million pounds of pesticides were applied to 11.1 million cotton acres in the United States. That indicates a national average of 6.43 pounds of pesticides applied per cotton acre. In California, an average of 12.8 pounds of pesticides per acre were applied to cotton in 1993 and 1994.
    It is widely documented that pesticides can affect the nervous, endocrine, immune, and reproductive systems, and that they pose heightened threats to infants, children, the unborn, and other subpopulations especially susceptible to toxic pollutants. The best documented and most visible environmental impact of pesticide use has been from toxic effects on wildlife, especially birds, fish, and other aquatic organisms. Aerial drift, runoff from treated fields, application error, food-chain contamination, and industrial accidents are the principal means by which pesticides end up damaging wildlife populations. Numerous studies have documented these effects. . . .
    Due to historically high pesticide use, insect resistance to commonly used pesticides is widespread in cotton. This means that pesticides are progressively less effective in controlling cotton pests. Cotton growers must then increase the amount and/or combinations of pesticides to get the same results.

Lately there has been strong interest in the cultivation of organic, chemical-free cotton. According to Agricola Partners, organic cotton is now grown in seventeen countries, and international demand for Earth-friendly cotton products has increased annually by about 25 percent since 1990. Some major marketers of cotton clothing, such as Patagonia, have even agreed to market organic cotton exclusively, in order to live up to their commitment to a more sustainable future.

    While these are positive developments, however, we must remember that cotton is also much more challenging to grow sustainably on a large agribusiness scale than industrial hemp. The lifting of hemp-growing restrictions would allow the marketplace to choose hemp (as well as organic cotton) over chemically sprayed cotton, and thus could bring about a great reduction in agricultural chemicals while keeping farmland in productive use.

Timber

Forests have evolved into complex biological systems that offer us—
in addition to natural beauty—food, fiber, oxygen, climate control, and
the ions that seed rain clouds. Thus, a grove of trees can represent a
grocery store, a lumber supply, a water fountain, a wildlife shelter, and
a heating and air-conditioning unit—all of the most aesthetically
pleasing design, and requiring no payment except that humanity respect the forest's ability to grow and evolve naturally.

Yet, time and again, humanity has failed to respect forestland, as
John Perlin describes in his classic book, *A Forest Journey*:

> The ruler at that time, Gilgamesh, wished to make a name for himself
> by building up his city. . . . Fortunately for Gilgamesh, a great pri-
> meval forest lay before him. That such vast tracts of timber grew near
> Southern Mesopotamia (Iran and Iraq) might seem a flight of fancy,
> considering the present barren condition of the land, but before the
> intrusion of civilizations an almost unbroken forest flourished in the
> hills and mountains surrounding the Fertile Crescent. . . . Gilgamesh's
> war against the forest has been repeated for generations, in every cor-
> ner of the globe, in order to supply building and fuel stocks needed for
> each civilization's continual material growth.

In our own time, for instance, Japanese companies and consumers
have been a major force in the destruction of forests to feed demand
for timber and paper products. Interestingly, most of the deforestation
has occurred abroad; Japan maintains its own forests (which cover 80
percent of this steep, mountainous land) in a sustainable, responsible
manner.

We in the United States have been overcutting forests in a nonsustainable manner, when we could be relying on annual crops to meet
the majority of our industrial fiber needs. When rains fall, the exposed
woodland soil is washed away to silt up our rivers. So the government
(we the taxpayers) hires wildlife biologists to conduct studies to determine why the salmon runs are diminishing. As the numbers of
salmon decrease, we pay more and more unemployment compensa-
tion to idled salmon fishermen and processors. Silted creeks and rivers
flood more easily, so we spend millions of dollars to remove dirt
washed down from the denuded mountains onto city streets and highways. This scenario is common throughout today's Pacific Northwest.

Mason Marsh, *Coos Bay World*

A mudslide that killed a motorist originated in a clear-cut above the road. More than 130 landslides occurred on Oregon Highway 38 between November 18th and December 3rd, 1996.

Yet, there are positive examples of forests being managed in a sustainable and profitable manner. For instance, Texada Logging, a company that owns 17,000 acres of private forest lands in British Columbia, has invested in European logging equipment with low-pressure tires to reduce soil compaction. The firm specializes in selective thinning of second growth forests. The company strives to remove no more than 30 percent of their forest stands, thus providing fish and fauna a wooded habitat instead of a clear-cut eroding hillside.

The need for more sustainable timber practices does not end at the logging site. Tree fibers contain 25 percent or more lignin (resinous plant glues). The traditional pulp and paper industry consumes large quantities of chemicals: sulfur compounds to separate lignin-rich tree fibers, and chlorine compounds to remove residual lignins during bleaching. Even utilizing modern wastewaster treatment systems, some of these compounds are released into rivers and oceans. In contrast, the stalks of annual crops such as hemp, kenaf, corn, and wheat have a low lignin content—less than 10 percent—in their fibers. Low-lignin crops such as hemp allow paper mills not only to consume less energy, but also to more easily avoid the use of sulfur and chlorine compounds. In fact, hemp can be bleached with environmentally friendly ozone or hydrogen peroxide. Such nonchlorine bleaching is a major feature of the zero-emissions paper "mini-mill" discussed later in this chapter.

Petroleum

The consequences of our global overreliance on petroleum and its derivatives is revealed to a staggering extent by the pollution of our air, soil, waterways, and oceans, as well as by rampant human health problems. Huge petrochemical processing factories break down oil into numerous materials for producing fuels, plastics, carpets, glues, resins, pesticides, and so on. Significant quantities of hazardous waste are produced in the process—as gases released into the atmosphere and as liquid sludge that is often injected into unlined wells 3,000 feet below ground.

Industry observers have estimated that, if the petroleum industry recycled and disposed of such waste in an environmentally responsible way, the total cost would be two to three times the industry's annual profits. If this were the case, our so-called cheap gasoline or plastic wrappers would be several times more expensive to buy, because society would not be subsidizing the oil industry's current pollution, which erodes both human and biological health. For example, people living near a belt of petrol refineries in Louisiana and Texas, nicknamed "Cancer Alley," are suffering increased health risks partly as a result of inaccurate or artificial pricing of petroleum and petrochemical products.

The book *Our Stolen Future* by Theo Coburn documents how common chemicals such as pesticides, dioxins, or additives in plastics actually disrupt hormonal systems in humans and wildlife. The Scandinavian countries are reporting a demasculinization of men due to high levels of toxic agricultural chemicals ("molecular garbage"), resulting in sperm-count losses of more than 50 percent along with other adverse physical and psychological changes. Some U.S. politicians, while promoting family values, are actually softening regulations that protect families from some toxins and pollutants.

As dismaying as our current problems seem, there are positive solutions on the horizon. A century ago, most industrial products were derived from plant matter; even the first plastics came from botanical sources such as cotton. Cellophane was named for cellulose, although today it is made primarily from oil.

This trend has been well researched by the Institute for Local Self-Reliance, based in Minnesota and the District of Columbia. In their report *The Carbohydrate Economy: Making Chemicals and Industrial Materials from Plant Matter*, authors David Morris and Irshad Ahmed explain:

In the 1920s, we moved away from an industrial economy based on living plants and toward an industrial economy based on dead, fossilized plant matter. The biological economy substitutes biochemicals for petrochemicals. In the last decade a combination of technological advances and environmental regulation has opened up new markets for industrial products derived from plant matter.

A biological economy must be based on sustainable cultivation and harvesting practices. The increased consumption of renewable materials must be matched by an increased commitment to agricultural techniques that preserve and enhance the quality of the soil.

Virtually all our non-energy, industrial product needs can be satisfied without expanding agricultural production. In fact, sufficient agricultural wastes exist to provide enough raw material to displace almost all petrochemicals.

A biological economy promises environmental and economic benefits. Environmentally, it benefits us by reducing our reliance on mining minerals and increasing our use of the stored chemical energy derived from sunlight, water, air, and trace minerals in the soil. Economically, a biological economy benefits us by adding jobs and creating industries in poorer rural areas.

If we are to return to a healthier, more life-sustaining ecosystem, we must break the petroleum stranglehold and develop a sound, biologically based economy in the twenty-first century. (Such a system is often described as a carbohydrate-based economy.) Besides reducing the use of petrochemicals by its very cultivation, many hemp products could replace petroleum-derived ones with truly biodegradable alternatives—disposable diapers, packaging materials, and even certain plastic resins, for example.

Geof G. Kime

A bumblebee feasts on sweet hemp flower pollen.

A CALL TO ACTION

Throughout a long history, the prestige and popularity of industrial hemp have risen and fallen to reflect technological innovations, or societal shifts brought about by war or world trade. With the need clear for a more sustainable economy, the time once again is ripe for hemp.

In the 1990s, hemp is seeing a resurgence in terms of acres planted and new products and businesses developed, yet many obstacles must still be overcome before hemp's full potential can be realized. These obstacles range from minimal seed-breeding activity to the lack of modern processing technologies to the counterculture image and legal hurdles that still bedevil true hemp. By clearing these hurdles, industrial hemp can deliver the many benefits that this plant has to offer. With increased investment in the design and development of appropriate machinery for harvesting, processing, and manufacturing, hemp products will become more cost-effective. And with increased education, perhaps the market will compel governments to lift unreasonable restrictions on the hemp industry.

Individuals or organizations that solve any of the following problems may discover any number of profitable and personally rewarding opportunities.

Seed Stock

Thanks to modern plant-breeding efforts, all the major agricultural crops—corn, wheat, rice, and cotton, for example—have seen significant increases in yield and other desired agronomic qualities since the end of World War II. By contrast, although hemp breeder Dr. Ivan Bócsa has done excellent work over the past three to four decades, hemp has been virtually ignored by other plant breeders around the world. In fact, hemp yield per acres is *less* today than it was in the 1930s.

Kentucky hemp commonly yielded five to eight tons of dry stalks per acre, compared to three to five tons per acre today. But this world-famous cultivar, which was developed over many plant generations, is now extinct. It will take many years to approach the yields of the 1930s. Yet, with committed focus from seed companies and agricultural breeders, there can be good progress in the coming decades.

There also has been very little work in the development of hemp varieties for seed oil or protein for human or animal consumption. A well-funded plant-breeding program should be able to increase the per-acre yield of hemp seed from its current seven hundred pounds to one thousand pounds or more.

In fact, some industry participants suggest the potential to obtain a yield of fifteen hundred to two thousand pounds of seed, and six to eight tons of dry stalk fiber per acre within five years, provided that expert botanists and breeders are funded to address this vital issue.

HempFlax's hemp harvester.

Harvesting and Retting Technology

As discussed in chapter 6, hemp-harvesting technology has stagnated since the 1930s, with few exceptions. Growers generally must resort to one of three choices: harvesting by hand, making do with antiquated equipment, or modifying modern equipment—sometimes with damage to machinery that cannot stand up to hemp's tough fibers.

Recently, HempFlax of the Netherlands has taken the lead in developing a specialized hemp harvester and a square-bale press. It would be most welcome to see other farm-equipment manufacturers

Malcolm MacKinnon

HempFlax fiber separation line at their Netherlands's factory.

follow HempFlax's example. Hemp growers also eagerly await innovative equipment for harvesting seeds, to better handle this crop's unusually tall (up to sixteen feet high) stalks. Additionally, alternatives to field-retting would free growers from the whims of the weather, which can ruin an entire harvest, while alternatives to water-retting would remedy odor and disposal problems. Ultrasonics and steam explosion are two possible alternatives being explored anew (the latter by several German research firms), after abandonment of these methods by Germany in the 1940s.

Fiber Separation

Hemp's sturdy fiber strands make possible a myriad of useful finished products, yet create unique challenges for processors. Like hay or straw, hemp stalks are bulky, and expensive to ship long distances. But the higher-value long fibers can be shipped farther, more cost-effectively. Thus it is essential that regional processors be located near hemp farms, as shown in chapter 5.

A key component of hemp processing is the fiber-separating process (decortication), whereby the long bast fibers are separated from the short core fibers. The development of such technology was crucial to hemp's 1930s resurgence in the United States (see chapter 2). As discussed in chapter 2, the fascinating machinery invented by Schlichten in 1917 reportedly separated hemp fibers without field-retting. Unfortunately, the crucial details of this innovation have been lost in time.

Most of the processing equipment being used in Europe today is modified flax equipment, much of it based on 1950s technology. The equipment available right now simply is not adequate to meet the rising demand for all kinds of hemp products.

Several companies, including HempFlax of the Netherlands, Hemcore of the United Kingdom, and Hempline of Canada, are working to improve hemp-decortication technologies in order to process field-retted stalks. When better machinery is developed, the cost of hemp will likely decrease, making its finished products more competitively priced, whether for paper, plastics, textiles, or building materials.

Manufacturing

Once hemp has gone through primary processing for fiber separation, it is ready for final processing, or manufacturing. As in all industries, this phase requires specialized equipment based on the end product.

Manufacturers must design intricate equipment and procedures in order to transform hemp fibers into finished products of consistent quality. We know that hemp was used in the 1930s for carpets, plastics, paper goods, building materials, auto-body panels, and many other applications. Yet that outmoded technology provides little help to manufacturers of the 1990s.

Questions that must be answered before industrial hemp can assume its rightful place in the world marketplace include:

- What is the most efficient and economical way to grow, harvest, and prepare the fibers?
- How should hemp be combined with other materials?
- How does the durability of hemp's end products come out in test comparisons with other materials?

Manufacturers around the globe are risking research and development funds to answer such questions in heretofore untried ways, but there is still tremendous opportunity for innovation on the frontiers of the hemp industry. Let's revisit two areas of particular challenge and promise: textiles and paper.

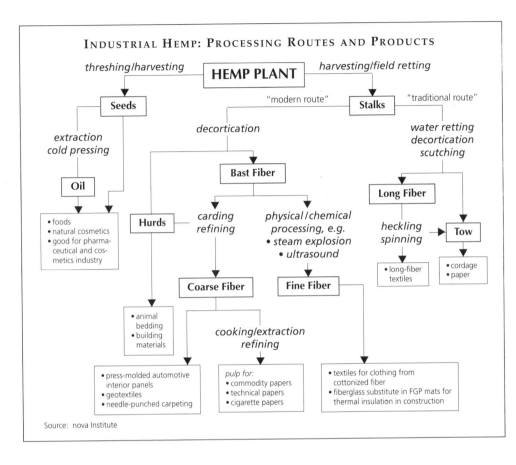

INDUSTRIAL HEMP: PROCESSING ROUTES AND PRODUCTS

Source: nova Institute

/ TEXTILES / Cotton is the reigning king of textile fibers in terms of consumer acceptance, ease of spinning, and infrastructure development. Not surprisingly, tens of billions of dollars have been invested in sophisticated spinning and processing equipment for cotton. On the other hand, hemp has seen less than $10 million invested in research and development for infrastructure since the 1940s. It is no wonder, then, that cotton equipment produces fabric faster, less expensively, and with fewer defects.

Even though hemp has significantly greater yields and lower growing costs per acre then cotton, hemp products are costlier because of the processing bottleneck. Whereas hemp fiber is much longer and stronger than cotton—a distinct advantage, from an end-product perspective—these same attributes make hemp very challenging to

Mackie International's Hempmack long fiber wet-spinning hemp textile machinery is produced at their Belfast, Ireland factory.

soften and spin. Today's cotton-spinning machinery cannot process the long, tough fibers. Moreover, although hemp can be spun on flax (linen) equipment, most of these machines are antiquated, with low output and high labor costs.

On a bright note, a much-improved hemp-spinning machine, called the HempMack, has been developed by Mackie International of Belfast, Northern Ireland. Mackie's engineering team designed the HempMack specifically to address the challenges of hemp-textile processing. It utilizes a wet-spinning process that offers labor savings over traditional machines and produces a stronger, better-quality yarn at higher speed. However, while the HempMack is a significant innovation for the hemp industry, it does not yet approach the production rates of cotton machinery. The traditional ring-spinning process for hemp, flax, and other bast fibers simply is slower, and thus more expensive, than rotor spinning for cotton.

German hemp technologists believe it unlikely that long-fiber textile spinning will be resurrected in their country for many years, because of the high costs involved in current labor-intensive methods. Therefore the Germans are focusing on alternative processes that will allow hemp to be spun on existing cotton equipment. Researchers and corporations have been working on ultrasonics and a steam-explosion process that weakens the fiber to create a hemp flock or "cotton." The shortened fiber can then be processed with conventional cotton-spinning equipment. This cottonization process was developed early in this century, but was abandoned for many years and has yet to achieve commercial success.

The good news is that industrial textile applications of hemp (often referred to as "technical textiles" and "nonwovens") look very promising from a commercial manufacturing perspective. Carpets, upholstery, and nonwoven medical and hygiene products (diapers, sanitary napkins, and the like) may be ideal uses for this durable and absorbent fiber. Such applications do not require the same fiber quality as garment textiles, and present far fewer processing challenges.

Even if all restrictions on cultivation were lifted today, most industry observers seem to think that many more years will pass before fine hemp textiles will be processed from raw hemp stalks on a large scale in North America. Yet in 1996, Giorgio Armani introduced a line of high-fashion hemp clothing. If industry giants such as Wrangler and Levi Strauss commit to the use of hemp, their decisions will accelerate the development of advanced hemp-spinning equipment. This in turn will open the hempen door for other clothing manufacturers, stimulating further advances in hemp-textile technology—a healthy cycle that could bring hemp very rapidly into the mainstream.

/ PAPER / Modern paper mills can cost from $500 million to $1 billion, and are designed to use wood pulp, which has different technical parameters than pulp from annual fibers such as hemp, kenaf, or agricultural straws (the leftover stalks of grain crops such as corn, wheat, and rice). Pulp for printing and fine writing papers must meet precise quality standards in order to run on complex, high-speed paper presses.

While major paper manufacturers are considering the use of hemp pulp, it will probably be at least another five years before they incorporate this into their mills. Is hemp paper therefore unlikely to be

La Chanvrière de l'Aube

Fine hemp paper made in France.

available for printing stationery and books such as this one? Fortunately, the prospects are not so gloomy. There is increasing interest in the idea of innovative "mini-mills" that can indeed process hemp pulp, while producing very low or zero emissions. Researchers such as Dr. Med Byrd at North Carolina State University's Department of Wood and Paper Science are developing plans for small paper mills that would process one hundred or more tons daily (as compared to three hundred to one thousand or more tons daily at larger mills). The mini-mills are designed to handle a combination of fibers, including hemp, kenaf, and agricultural straws, as well as recycled fibers.

These proposed mini-mills represent a major breakthrough in paper manufacturing. Some wood-pulping processes release noxious sulfur and chlorine compounds into our air and waterways. The goal of a zero-emissions facility is to transform any pulping by-products into valuable resources, such as fertilizers for future crops. If hemp can be pulped using fewer and more benign chemicals, it is possible that the by-products (liquors, sludge) could be spread back onto the fields or converted into other useful products. Such a closed-loop system is a prime example of industrial ecology. High-quality paper products can also be made from hemp's strong medium-length fibers, which are an abundant by-product of the fiber separation process. However, converting hemp-fiber pulp into lower-grade paper products (such as cups, napkins, egg cartons, and fast-food clamshells) is a much less technically demanding process than milling high-grade printing papers, and may constitute worthwhile initial applications for hemp pulp. Unfortunately, such products are often limited to commodity pricing, so the higher raw-material costs for processing hemp could be a disadvantage. There also is considerable interest in transforming the short core fibers into paper products, yet further research is needed. One possible low-tech approach involves molding the short fibers into clamshells and other packaging materials. The Ukrainian Pulp and Paper Institute in Kiev, Ukraine, is developing a process that can produce paper pulp from the entire hemp stalk—without separating the long and short fibers. In 1997, the U.S. Department of Agriculture's Forest Products Laboratory in Madison, Wisconsin, produced a report entitled "Market Analysis for Hemp Fiber as a Feed Stock for Papermaking," which said in part:

> In contrast to the past utilization of hemp, it is essential that the whole plant be used to make paper and not just the long bast fibers. . . . A chemical analysis of the hurds suggests that they are 55 percent cellulose and 25 percent lignin, which is similar to many hardwoods. . . . An analysis of the bast fibers shows that they are composed of 70 percent cellulose and 8 percent lignin. . . . With the tightening of the domestic wood-chip supply, there is a strong upward price pressure. . . . This simple analysis shows that Wisconsin farmers could profitably produce hemp, and that they could meet the fiber demand in the state if 530,000 acres were planted in hemp.

Tree pulp today is extremely and artificially inexpensive, despite its diminishing supply, due to a combination of the low value placed on forests and the subsidies that the timber industry receives from governments. Obviously, until hemp fiber becomes available in larger quantities at competitive prices, the pulp and paper industry will expend neither the funds nor effort to retrofit large factories for processing hemp. And, of course, without a reliable market for raw hemp fiber, farmers will be reluctant to grow hemp. It remains to be seen which will come first: the chicken or the egg.

The Law

There have been no permits given in the United States to grow hemp since the 1950s. DEA chief Barry McCaffrey wrote on June 30, 1997, to Kentucky governor Paul F. Patton:

> The end result of legalizing hemp production might well be de facto legalization of the cultivation of marijuana. Hemp and marijuana are the same plant, the seeds are the same, the seedlings are the same, and in many instances the mature plants look the same.
>
> The facts are that hemp production has extremely limited economic potential. Hemp production does not yet appear to be a useful new crop for hard-pressed farmers. Certainly we should consider any new credible evidence. However, the current facts do not support hemp cultivation as an economically viable option for U.S. agriculture.

In twenty-nine countries the cultivation of hemp is legal and marijuana is not—so the DEA's "de facto legislation of marijuana" claim is not based on the facts. Additionally, when was the DEA empowered to pass judgment on the economic viability of agricultural crops? Should we not let the marketplace determine if hemp is economically worthwhile to grow, process, or manufacture?

It is obvious that the DEA is not friendly to the hemp industry. The best solution, as mentioned in previous chapters, is to transfer the primary regulatory responsibility for hemp from the DEA to the USDA. This can be accomplished by an executive order signed by the president of the United States and does not require any new federal legislation. Yet it is unlikely that this will occur until significant political pressure is applied to government at both the state and federal levels. Eventually, legal suits challenging the current outmoded laws may bring forth change. Another possible solution would be for each individual state to pass a bill permitting the growing of industrial hemp. Of course, the DEA could attempt to stop such state-sanctioned plantings, yet that could lead to a highly publicized federal-versus-state-government showdown on the hemp issue. It will be exciting to watch how these developments unfold in the coming years.

Education

Even though hemp is receiving considerable media coverage around the world, most people still don't comprehend the important role that this fiber will play in the twenty-first century. The average citizen knows but a small piece of the truth about industrial hemp, along with a good deal of misinformation.

I receive weekly inquiries from students and teachers nationwide who want reliable information about industrial hemp for use in school projects. If hemp is mentioned at all in history textbooks that cover the fifteenth through the twentieth centuries, it occurs within a brief sentence or two about rope. Standard histories overlook food, feed, sails, canvas, paint, paper, plastics, and lubricants, and the fact that hemp was the world's foremost agricultural crop during the eighteenth century. Of course, the most widespread misunderstanding is that hemp is the same variety as marijuana. The history books fail to state the full story of industrial hemp, including the fact that it was expressly distinguished from marijuana and protected as a fiber and seed crop in the United States congressional legislation of 1937 and 1946.

As proponents of industrial hemp we must first inform ourselves, discerning the truth amidst the hype. Next, we must educate farmers, policymakers, businesspeople, and consumers to understand that hemp is a versatile, renewable raw material—one well worth incorporating into a more global sustainable future.

Shocks of hemp in Kentucky, circa 1930s.

How Hemp Could Evolve in the United States

1988 *Counterculture*—Independent individuals increasingly bang the drum like town idiots, shouting a ridiculous story that no one wants to hear or believe.

1991 *Entrepreneurs*—A handful of enterprising adventurers market hemp products at high prices to true believers.

1994 *Agents of Change*—A wave of bridge-builders links distinctly and fundamentally different groups within society.

1996 *Vanguard Organizations*—Leaders carry the torch into their respective organizations and industries, withstanding severe criticism as changes gradually come to previously inflexible, hierarchical systems.

1998 *The Official Stamp of Approval*—Numerous government leaders and large corporations publicly endorse industrial hemp.

1999 *Agricultural Development*—The agricultural community moves full-speed ahead to research and develop a domestic infrastructure for the hemp industry.

2000 *A Technology Shift*—Harvesting, processing, and manufacturing roadblocks are significantly removed, opening the way for economically competitive products.

2001 *Regional Hot Spots*—Some regions of the country emerge where hemp activity is driving economic growth and where citizens purchase and prefer locally grown and processed hemp goods.

2004 *Major Markets*—As supplies of raw materials increase and prices fall, major marketers move strongly into the marketplace to meet rising consumer demand.

2010 *The Hemp Industry Reaches Maturity*—The industry has gained a large market share in dozens of huge commercial sectors, and hemp products are now commonly found at home centers, shopping malls, and supermarkets.

What You Can Do

In addition to spreading the word about industrial hemp, each one of us can demonstrate our advocacy through personal action. By requesting and purchasing hemp products, we are voting with our dollars and supporting the expansion of the hemp industry. Help multiply the number of industrial-hemp supporters by sending a copy of this book to a friend, relative, or civic leader. Consider writing an informative letter to your elected officials, government-agency representatives, or newspaper editors, or making a factual presentation to a community group.

Changing the Dream

In thinking about how real change is brought about, I'm reminded of an inspiring story that I heard at a conference, told by John Perkins, a retired World Bank executive. Perkins had traveled deep into the Amazon to learn how the indigenous people there live in balance with the Earth. He was feeling some guilt about how the unintended consequences of our Western affluence have destroyed the forest habitats of many native tribes. His guide in the rain forest was an older man who was a shaman in his tribe. At the end of their time together, just before this elder tribesman paddled away in his canoe, the banker asked the shaman what he, the Westerner, as just one person, could do to change things for the better. The shaman thought for a moment, then replied, "Just change the dream."

Many of us now realize that we must awaken from the old dream of a comfortable lifestyle made possible by nonsustainable consumption. Let's all heed the advice of the shaman and "change the dream."

Hemp Essentials

Hemp Essentials is a cottage industry owned and operated by Carol Miller in northern California's rugged coastal mountains. Before starting her home business, this high-school teacher had for many years operated a bath-products enterprise called Herbal Delight. Miller also had cooked with hemp seed and its oil for twenty years.

As Hemp Essentials has grown, many people have encouraged Miller to expand her business beyond the production limits of her home. However, she has decided to keep managing Hemp Essentials as a cottage industry. At home, she can raise her six children and three grandchildren while continuing to offer employment to family members.

How has Miller been able to keep her growing business more or less under one roof? Flexibility and creativity have been the key:

> We have expanded our production capacity by adding storage, including solar cold-storage for the lotions and oils, and soap-drying racks in a loft over a wood-burning stove. Our greenhouse, which nurtures baby melons and cucumbers in the winter, becomes a solar dryer for soap in the summer and fall, while the inside racks accommodate dried flowers for my gift boxes and herbs for the healing salves, all organically grown here in our gardens.

Miller understands the expanding appeal of Hemp Essentials. She says:

> Hemp products are in demand because they represent progress toward a sustainable-yield culture. This country has been strip-mined of its resources, and the American public knows that jobs are leaving the mainland, creating unemployment and growing poverty. Hemp can heal the damage done to our economy—just as it can heal the damaged soil—just as it heals our skin, cardiovascular systems, and nervous and immune systems. Hemp heals, and people know that such healing is needed—and fast.
>
> Every time a new customer uses a hemp product, they experience the healing and know that they are participating in a larger healing. When I photocopy tests and handouts for my class on hemp paper, the teenagers show respect that is worth every penny of the extra cost! In many ways, it is young people and their concern for the future that has fueled the growth of the industry.

Hemp Essentials
P.O. Box 151
Cazadero, CA 95421
Telephone: (707) 847-3642

HEMP RESOURCES

Accessories

Artisan Weavers
P.O. Box 307
Middlebury, VT 05753
Tel: (802) 388-6856
Fax: (802) 388-2150

The Hemp Club, Inc.
3418 A Park Avenue
Montreal, QC H2X 2H5
Canada
Tel: (514) 845-4993
Fax: (514) 845-4687
plss57@prodigy.com

Hempstead Company
2060 Placentia, #B-2
Costa Mesa, CA 92627
Tel: (800) 284-HEMP
  or (714) 650-8325
Fax: (714) 650-5853
www.hempstead.com

Lost Harvest
135 McDonough St., #23
Portsmouth, NH 03801
Tel: (800) 325-4367
Fax: (603) 431-1489

The Natural Order, Inc.
P.O. Box 73 Stn. P
Toronto, Ontario M5S 2P6
Canada
Tel: (416) 588-4209
Fax: (416) 588-4210

Steba Ltd.
Kaszalo Str. 139
Budapest, Hungary, 1173
Tel: 36-1-257-2745
Fax: 36-1-256-9802
steba@hungary.net

Terra Pax
2145 Park Ave., Ste. 9
Chico, CA 95928
Tel: (916) 342-9282
Fax: (916) 342-3730
www.terrapax.com

Apparel

AdventureSmiths
P.O. Box 50333
Eugene, OR 97405-0977
Tel: (541) 343-9924
Fax: (541) 343-6103
advntrsm@efn.org

All Points East
P.O. Box 221776
Carmel, CA 93922
Tel: (408) 655-4367

Colour Connection
Herzo Base 1595
Herzogenaurach 91074 Germany
Tel: 49-9132-1790
Fax: 49-9132-4176

Ecolution (see Distributors)

Giggling Piglet Co-op
Home of Hempenware
2103 Harrison NW, Ste. 2756
Olympia, WA 98502
Tel: (306) 705-3804
www.olywa.net/uncleweed

Headcase Hemp Hats
150 Bay Street
Jersey City, NY 07302
Tel: (201) 420-5900
Fax: (201) 420-7101
hemphats@aol.com
www.headcase.com

Hemptown
BC Hemp Co. Inc., Ste. 203
196 West Third Avenue
Vancouver, BC, Canada V5Y 1E9
Tel: (604) 708-5051
Fax: (604) 708-5052
hemptown@uniserve.com
www.iceonline.com/home/
   hemptown/hemptown.html

Home Grown Hats
P.O. Box 1083
Redway, CA 95560
Tel: (888) ECO-HATS
Tel/Fax: (707) 986-7713

Island Hemp Wear
P.O. Box 690
Kekaha, HI 96752
Tel: (808) 337-1487
Fax: (808) 337-1917
hemp@hawaiian.net
www.hawaiian.net/~hemp

Jus Naturale
5020 N.E. 180th Street
Seattle, WA 98155
Tel: (206) 365-4375
Fax: (206) 363-9310

Labyrinth
463 Haight Street
San Francisco, CA 94117
Tel: (415) 552-3082
Fax: (415) 552-3083
labrnth@sirius.com

Mendocino Hemp Co.
32101 Middle Ridge Road
Albion, CA 95410
Tel: (707) 937-5529
Fax: (707) 937-5768

Of The Earth
916 W. Broadway #749
Vancouver, BC V5Z 1K7 Canada
Tel: (604) 878-1268
Fax: (604) 871-0082
sales@oftheearth.com
www.oftheearth.com

Tribal Fiber (see Housewares)

Two Star Dog
1370 10th Street
Berkeley, CA 94710
Tel: (510) 525-1100
Fax: (510) 525-8602
www.twostardog.com

Associations and Organizations

Agricultural Hemp Association Voter
P.O. Box 8671
Denver, CO 80201
Tel: (303) 298-9414
Fax: (303) 280-0130
ahavoter@aol.com

Australian Hemp Industries Association
P.O. Box 236
New Lambton, New Castle
Australia 2305
Tel: 61-49-55-6666
Fax: 61-49-55-6655
Austhemp@hunterlink.net.au

Austrian Hemp Institute
Duererg. 3/4
Wien, Austria A-1060
Tel: 431-586-9429
Fax: 431-586-9448
oekoforum@magnet.at

Center of the FAO Network on
  Flax & Hemp
c/o European Co-operative Research
  Network on Flax
Institute of Natural Fibres
Ul. Wojska Polskiego 71B
Poznan, Poland 60-630
Tel: (48 618) 480-061
Fax: (48 618) 417-830
SLAWA@ruby.poz.edu.pl

Chlorine Free Products Association
102 N. Hubbard Street
Algonquin, IL 60102
Tel: (847) 658-6104
Fax: (847) 658-3152
CFPA@ibm.net

Danish Hemp Society
Pilestraede 8b, DK-1112
Copenhagen K, Denmark
Tel: 45-3313-2328
Fax: 45-3313-2329
npfd@inet.uni-c.dk

Galloway Fields Co., Inc.
Mid-South Fiber Network
7390 Raleigh Lagrange Road
Cordova, TN 38018
Tel: (901) 757-1777
Fax: (901) 757-9568

Hemp Industries Association (HIA)
P.O. Box 1080
Occidental, CA 95465
Tel: (707) 874-3648
Fax: (707) 874-1104
info@thehia.org
www.thehia.org

Hemp Industries Marketing Board of
  New Zealand Ltd.
P.O. Box 11-015
Wellington, New Zealand
Tel: (800) 436769 or 64-4-499-8876
Fax: 64-4-499-8875
hemp@ihug.co.nz
www.hempboard.co.nz

Institute for Local Self Reliance
1313 5th Street SE, Ste. 303
Minneapolis, MN 55414-4513
Tel: (612) 379-3815
Fax: (612) 379-3920
dmorris@ilsr.org
www.ilsr.org

Institute for Natural Fibers
Ul. Wojska Polskiego 71B
Poznan, Poland 60-630
Tel: (48 618) 480-061
Fax: (48 618) 417-830
dgucia@iwn.inf.poznan.pl

International Hemp Association (IHA)
Postbus 75007
1070 AA Amsterdam, Netherlands
Tel/Fax: 31-20-6188758
iha@euronet.nl

International Kenaf Association
101 Depot Street
Ledonia, TX 75449
Tel: (903) 367-7216
Fax: (903) 367-7060

Kentucky Hemp Growers Cooperative
  Association, Inc.
P.O. Box 8395
Lexington, KY 40533
Tel: (606) 252-8954
WdavidS100@aol.com

Kentucky Hemp Museum & Library
P.O. Box 8551
Lexington, KY 40533
Tel: (502) 465-7672 or (502) 692-977

Natural Step
4000 Bridgeway, Ste 102
Sausalito, CA 94965
Tel: (415) 332-9394
Fax: (415) 332-9395
www.emis.com/tns

New Uses Council
P.O. Box 144
Jamestown, RI 02835-0144
Tel: (401) 423-0862
Fax: (401) 423-0862
jharsh@rof.com

New Zealand Hemp Industries
    Association, Inc.
P.O. Box 38-392
Howick, Auckland, New Zealand
Tel: 025-514-229-04-477-4819
nzhemp@es.co.nz

North American Industrial Hemp
    Council
P.O. Box 259329
Madison, WI 53725-9329
Tel: (606) 224-5135
Fax: (606) 224-5110
sholtea@wheel.datcp.state.wi.us
www.naihc.com

nova Institute (see Consulting)

Organic Fiber Council
P.O. Box 72424
Davis, CA 95617
Tel: 916-750-2265
Fax: 916-756-7188
agaccess@davis.com

Pesticide Action Network
116 New Montgomery #810
San Francisco, CA 94105
Tel: (415) 541-9140
Fax: (415) 541-9253
panna@panna.org

Sustainable Cotton Project
6176 Old Olive Hwy
Oroville, CA 95966
Tel: (916) 589-2686

USDA AARC
Washington, DC 20250-0401
Tel: (202) 690-1633
Fax: (202) 690-1655
www.usda.gov/arc/

Body Care

All Around the World Hemp
154 Trescony Street
Santa Cruz, CA 95060
Tel: (800) 449-4945
Fax: (408) 426-4263
hemppoet@worldhemp.com
www.worldhemp.com

Artha
P.O. Box 20154
Oakland, CA 94620
Tel/Fax: (510) 420-0696

Hemp Essentials
P.O. Box 151
Cazadero, CA 95421
Tel:(707) 847-3642
http://hempworld.com/hempess.html

HempGarden
Pilgerstr. 76
90429 Overath-Marialinden
Germany
Tel: 49-2206-869693

Hempola
R.R. #1
Port Severn, Ontario L0K 1S0
Canada
Tel: (800) 240-9215 or
(905) 678-1066
Fax: (905) 678-6036
constr@interlog.com
www.hempola.com

High Desert Trading Co.
P.O. Box 247
Elfrida, AZ 85610
Tel: (800) 603-6601
Fax: (520) 642-9385
highdeserthemp@theriver.com

Jason's Natural Cosmetics
8468 Warner Dr.
Culver City, CA 90232
Tel: (310) 838-7543
Fax: (310) 838-9274
jnp@jason-natural.com
www.jason-natural.com

The Merry Hempsters
P.O. Box 1301
Eugene, OR 97440
Tel: (541) 345-9317
Fax: (541) 345-0910
merryhmp@pacinfo.com

Ohio Hempery (see Distributors)

Two Star Dog (see Apparel)

Building Materials

Ag Innovation (see Seeds and Seed Cultivars)

La Chanvrière de l'Aube
Rue General de Gaulle
10 200 Sar-sur-Aube, France
Tel: 33-3-2592-3192
Fax: 33-3-2527-3548
chanvriere@marisy.fr
www.marisy.fr/chanvriere

RoHemp (see Distributors)

Carpets

Dharmic Enterprises
P.O. Box 588
Topanga, CA 90290
Tel: (310) 455-4241
Fax: (310) 455-1046
lunadahrm@earthlink.com

Interface Inc.
100 Chastain Center Blvd., Ste. 165
Kennesaw, GA 30144
Tel: (770) 421-9555
Fax: (770) 424-1888

Naturlich
7120 Keating Avenue
Sebastopol, CA 95472
Tel: (707) 829-3959
Fax: (707) 829-1774
www.monitor.net/~nathome/

Ortex (see Cordage)

Catalogs

Frankel Bros Hemp Outfitters
(see Retail)

Ohio Hempery (see Distributors)

Oxford Hemp Exchange
(see Distributors)

Real Goods Trading Corp.
555 Leslie Street
Ukiah, CA 95482
Tel: (800) 762-7325
Fax: (707) 468-9486

realgoods@realgoods.com
www.realgoods.com

Consulting

Danish Hemp Society (see Associations and Organizations)

HEMPTECH
P.O. Box 1716
Sebastopol, CA 95473
Tel: (707) 823-2800
Fax: (707) 823-2424
john@hemptech.com
www.hemptech.com

nova Institute
Thielstr. 33
50354 Hurth, Germany
Tel: 49-2233-978374
Fax: 49-2233-978369
100675.1134@compuserve.com

Cordage

American Hemp Mercantile
124 Williams Avenue So.
Renton, WA 98055
Tel: (206) 204-877
Fax: (206) 204-9606

Ortex
19631 Vision Drive
Topanga, CA 90290
Tel: (310) 455-9431
Fax: (310) 455-9435
agi@jovanet.com

Distributors

American Hemp Mercantile
124 Williams Avenue So.
Renton, WA 98055
Tel: (206) 204-877
Fax: (206) 204-9606

Australian Hemp Products
P.O. Box 236
New Lambton, New Castle
Australia 2305
Tel: 61-49-55-6666
Fax: 61-49-55-6655
Austhemp@hunterlink.net.au

Ecolution
2812-E Merrilee Drive
Fairfax, VA 22031
Tel: (888) ECO-HEMP or
(703) 207-9001
Fax: (703) 560-1175
orders@ecolution.com
www.ecolution.com

Frankel Bros Hemp Outfitters
   (see Retail)

Ohio Hempery
P.O. Box 18
Guysville, OH 45735
Tel: (800) BUY-HEMP or
(614) 662-4367
Fax: (614) 662-6446
hempery@hempery.com
www.hempery.com

Oxford Hemp Exchange
2725 N. Westwood Blvd., Ste. 7
Popular Bluff, MO 63901
Tel: (573) 785-8711
Fax: (573) 785-3059
oxhemp@pbmo.net
www.pbmo.net/oxhemp

RoHemp GmbH
Wallstr. 36
Furstenfeld, Austria 8280
Tel: 43-3382-52300
Fax: 43-3382-52301
rohemp@styria.com
www.hempworld.com/rohemp [English]
www.hanfnet.de/rohemp.html [German]

Schermerhorn
12922 Florence Avenue
Santa Fe Springs, CA 90670
Tel: (800) 932-9395
Fax: (562) 946-4627
sbcosfs@msn.com

Steba Ltd. (see Accessories)

Victory Sales Group
2319 N. 45th Street, #164
Seattle, WA 98103
Tel: (800) 436-7783 or
(206) 322-0903
Fax: (206) 322-5664

Food

Frederick Brewing Co.
4607 Wedgewood Blvd.
Frederick, MD 21707
Tel: (301) 694-7899
Fax: (301) 694-2971
kevin@fred.net
www.fredbrew.com

Hempfields Natural Goods
11798 Detroit Avenue
Cleveland, OH 44107
Tel: (888) HEMP-JAVA or
(216) 226-0660
Fax: (216) 226-2229
www.hempfields.com

Hempola (see Body Care)

Hungry Bear Hemp Foods
P.O. Box 12175
Eugene, OR 97440
Tel: (888) MMM-HEMP or
(541) 345-5216
Fax: (541) 302-1488
eathemp@efn.org
www.efn.org/gordon_k/HungryBear.
   html

New Earth Ltd.
P.O. Box 204, Barnet
London EN5 1EP England
Tel: 10-44-0-378-559-242

Rella Good Cheese Company
P.O. Box 5020
Santa Rosa, CA 95402
Tel: (707) 576-7050
Fax: (707) 576-7116
richard@rella.com
www.rella.com

Footwear

Casa Natura
Weberstr. 66-HH
60318 Frankfurt/Main Germany
Tel/Fax: 49-69-552735

Deep E Company
322 N.W. Fifth Ave., Ste. 207
Portland, OR 97209
Tel: (888) 233-3373
Fax: (503) 299-6287
deepeco@aol.com

Deja Shoe
28725 S.W. Boones Ferry Road
Wilsonville, OR 97070
Tel: (503) 682-8814
Fax: (503) 682-9221
10346.1650@compuserve.com

Eco Dragon
P.O. Box 374
Guilford, CT 06437
Tel: (203) 245-3400
Fax: (203) 245-3405
Ecodragon@aol.com

Furnishings

Crown City Mattress
250 S. San Gabriel Blvd.
San Gabriel, CA 91776
Tel: (818) 796-9101
Fax: (818) 287-6819
crown1@mail.idt.com

Rio Rockers
2120 N. Pacific Avenue #37
Santa Cruz, CA 95060
Tel: (800) 746-2656 or
(408) 426-0265
Fax: (408) 469-3011

Whittle Works
92 Bethlehem Pike
Philadelphia, PA 19118
Tel: (215) 247-1996
Fax: (215) 247-5112

Government Licensing

Bureau of Drug Surveillance
122 Bank Street, 3rd Floor
Ottawa, Ontario K1A 1B9
Canada
Tel: (613) 954-6524
Fax: (613) 952-7738

HOME OFFICE
SE Region Drugs Inspectorate
50 Queen Anne's Gate, London SW1H
9AT
Tel: 0171-273-3856
Fax: 0171-273-2671

USDEA
Room W12058
700 Army Navy Drive
Arlington, VA 22202
Tel: (202) 307-7927
Fax: (202) 307-4502

Housewares

Earth Runnings
P.O. Box 3027
Taos, NM 87571
Tel: (505) 758-5703
Fax: (505) 737-0679
earthrun@laplaza.org
http://ecomall.com/biz/earth.htm

Fremont Hemp Company (see Retail)

Grassworks
P.O. Box 939
Captain Cook, HI 96704
Tel: (800) 472-7795 or
(808) 328-8803

Of The Earth (see Apparel)

Tribal Fiber, Inc.
P.O. Box 19755
Boulder, CO 80308
Tel: (303) 415-0478
Fax: (303) 440-6006
tribalfiber@earthlink.net

Internet Sites

www.hemptech.com

www.hempseed.com

www.naihc.org

www.thehia.org

Machinery

Ecco-Gleittechnik GmbH
Salzsteinstr. 4
82402 Seeshaupt, Germany
Tel: 49-8801-97-0
Fax: 49-8801-97-30

HempFlax B.V. (see Processors)

Mackie International
P.O. Box 4417
Dalton, GA 30719
Tel: (800) 330-8567
Fax: (706) 278-2697

Oil

All Around the World Hemp
(see Body Care)

Hempola (see Body Care)
Ohio Hempery (see Distributors)

Spectrum Naturals
133 Copeland Street
Petaluma, CA 94952
Tel: (707) 778-8900
Fax: (707) 765-1026
spectrumnaturals@.netdex.com
www.spectrumc.com

Paper

Crane Paper
30 South Street
Dalton, MA 01226
Tel: 413-684-2600
Fax: 413-684-0726

EcoSource Paper, Inc.
111-1841 Oak Bay Avenue
Victoria, BC V8R 3N6 1C4
Canada
Tel: (800) 665-6944 or (250) 595-HEMP
Fax: (205) 370-1150
ecodette@island.net.com
www.islandnet.com/~ecodette/ecosource.htm

The Friendly Chameleon
223 Sumac Street
Philadelphia, PA 19128
Tel: (800) 717-8242 or (215) 487-3317
Fax: (215) 508-1699
chameleon@igc.apc.org

Hemcore Ltd. (see Processors)

Highland Hemp
P.O. Box 1
Strathblane, G63 9EA Scotland
Tel/Fax: 44-1360-771153
chris@hhemp.demon.co.uk

Living Tree Paper
1430 Wilamette St., Ste. 367
Eugene, OR 97401
Tel: (800) 309-2974 or
(541) 342-2974
Fax: (541) 687-7744
livingtree@dnsi.net
www.hempnet.com/livingtree

Schneidersohne Papier GmbH + CokG
Benzstr. 3
65779 Kelkheim, Germany
Tel: 49-6195-801-0

Steba Ltd. (see Accessories)

Plastics

Zellform
Reiderstr. 3
Taiskirchen, Austria A-4753
Tel: 43-77-64-7722
Fax: 43-77-64-

Processors

Badische Naturfaseraufbereitung (BaFa)
Am fuchseck 1
76316 Malsch, Germany
Tel: 49-721-891300
Fax: 49-721-891301

Consolidated Growers and Processors
Inc.
P.O. Box 2228
Monterey, CA 93942-2228
Tel: (888) 333-8CGP
Fax: (888) 999-8GCP
info@congrowpro.com
www.congrowpro.com

Hemcore Ltd.
Station Road, Felsted
Great Dunmow, Essex
England CM6 3HL
Tel: 441-371-820066
Fax: 441-371-820069
stuart@hemcore.demon.co.uk

HempFlax B.V.
P.O. Box 142
9665 ZJ Oude Pekela
The Netherlands
Tel: 31-597-615-516
Fax: 31-597-615-951
info@hempflax.com
www.hempflax.com

Hempline Inc.
10 Gibson Drive
Tillsonburg, Ontario N4G 5G5
Canada
Tel: (519) 434-3684
Fax: (519) 434-6663
fiber@hempline.com
www.hempline.com

Kenex Ltd.
R.R. #1, Pain Court
Ontario, Canada N0P 1Z0
Tel: (519) 352-2968
Fax: (519) 352-6667
www.kenex.org

La Chanvrière de l'Aube (see Building
Materials)

Rohemp GmbH (see Distributors)

Publications

*Commercial Hemp: The Trade Journal
for Our Growing Industry*
909 Windermere Street
Vancouver, B.C. V5K 4J6
Canada
Tel: (604) 258-7171
Fax: (604) 258-7144
www.wisenoble.com

Hanf!
Das Grosse Journal Fur Hanf-Kultur
Postfach 7
79233 Vogtsburg, Germany
Tel: 49-7662-911990
Fax: 49-7662-911995

Hanfmagazine
Durergasse 3/4
1060 Wien/Vienna, Austria
Tel: 43-1-586-9429
oekoforum@magnet.at

HempWorld
P.O. Box 550
Forestville, CA 95436
Tel: (707) 887-7508
Fax: (707) 887-7639
mari@hempworld.com
www.hempworld.com

Hemp Magazine
1533 Westhiemer
Houston, TX 77006
Tel: (713) 523-3199
Fax: (713) 528-4367
hempmag@nol.net

Hemp Times Magazine
235 Park Ave. S.
New York, NY 10003
Tel: (212) 260-0200
Fax: (212) 475-7684

*Journal of the International Hemp
Association* (see Associations and
Organizations)

Les echos du Chanvre
61 Avenue Jean Jaures
69007 Lyon, France

Research

Australian Hemp Resource and
Manufacture (see Seeds and Seed
Cultivars)

Consolidated Growers & Processors (see
Processors)

Danish Hemp Society (see Associations
and Organizations)

DLR Institut fur Strukturmechanik
Lilienthalplatz 7
38108 Braunschwieg, Germany
Tel: 49-531-295-2310
Fax: 49-531-295-2838
hermann@dv.bs.dlr.de

Institute for Applied Research
Kai Nebel
Alteburgstr. 150
72762 Reutlingen, Germany
Tel: 49-7121-271-536
Fax: 49-7121-271-537
kai.Nebel@IH.Reutlingen.de

Institute of Natural Fibres (see
Associations and Organizations)

Dr. Paul G. Mahlberg
Indiana University
Department of Biology
Jordan Hall
Bloomington, IN 47405
Tel: (812) 855-5980
Fax: (812) 855-6705

nova Institute (see Consulting)

Retail

2000 BC., Inc.
8260 Melrose Avenue
Los Angeles, CA 90046
Tel: (213) 782-0760
Fax: (213) 655-2091
www.2000bc.com

American Hemp Mercantile
401 Broadway Market East
Seattle, WA 98102
Tel: (206) 323-6778

Atlanta Hemp
376 Oakdale Road
Atlanta, GA 30307
Tel: (404) 524-4367
Fax: (404) 370-9677
link@mindspring.com

Austin Hemp Company
3401 Guadalupe Street
Austin, TX 78705
Tel: (512) 302-1088
Fax: (512) 302-1094
kat@hemponline.com
www.hemponline.com

Blue Sativa
439 Frandor Avenue #38
Lansing, MI 48912
Tel: (888) 505-HEMP or
(517) 336-9336
Fax: (517) 336-9811
www.avwd.com/bluesativa

British Hemp Stores
76 Colston Street
Bristol, BS1 5BB England
Tel: (44) 117-9298-371
Fax: (44) 117-9238-326

California Organic Cotton Company
1247 3rd Street
Santa Monica, CA 90401
Tel: (310) 395-1044
Fax: (310) 395-2014

Cha! – Exclusively Hemp
7101 E. Stetson
Scottsdale, AZ 85251
Tel: (602) 675-0287
chaxh@aol.com

Frankel Bros. Hemp Outfitters
P.O. Box 26628
San Francisco, CA 94126
Tel: (888) 420-FBHO
Fax: (415) 753-0998
dave@frankelbros.com
bob@frankelbros.com
http://FrankelBros.com

Fremont Hemp Company
3526-C Fremont Place N.
Seattle, WA 98103
Tel: (206) 632-HEMP
Fax: (206) 632-1266
www.fremonthemp

Friendly Stranger
226 Queen Street W.
Toronto, Ontario M5V 126
Canada
Tel: (416) 591-1570
Fax: (416) 591-8438
friend@friendlystranger.com
www.friendlystranger.com

Green Lands
Utrechtsestraat 26, 10A Amsterdam
1000 BR Amsterdam, Netherlands
Tel: 31-20-625-1100
Fax: 31-20-422-1125

Hanf Haus GmbH
Waldemarstr. 33
10999 Berlin
Tel: 49-30-6149884
Fax: 49-30-6149911

Hemp in the Hollow
640 S. Coast Hwy. #2A
Laguna Beach, CA 92651
Tel: (714) 494-3070
Fax: (714) 494-1065
hemphollow@aol.com

Hemporium New Zealand, Ltd
151 Cuba Street
Wellington, New Zealand
Tel: (800) GET-HEMP
Tel/Fax: 64-4-385-2907
hemp@hemporium.co.nz

Hemp Shak
240 W. Foothill Blvd.
Claremont, CA 91711
Tel: (909) 398-1041
hempshak@erthlink.net

Hemp Shak
570 Higuera, Unit #18
San Luis Obispo, CA 93401
Tel: (805) 543-0760
hempshak@earthlink.net

Oxford Hemp Exchange (see Distributors)

Planet Hemp
423 Broome Street, SoHo
New York, NY 10013
Tel: (212) 965-0500
Fax: (212) 965-0873
www.planethemp.com

Santa Barbara Hemp Company
741 De La Guerra Plaza
Santa Barbara, CA 93101
Tel: (805) 965-7170
sbhempco@gte.net

Third Stone Hemp
703 West Lake Street
Minneapolis, MN 55408
Tel: (612) 825-6120
Fax: (612) 824-8219
www.harmonypark.com

Two Star Dog (see Apparel)

Washington Hemp Mercantile
217 S. First Street
Mt. Vernon, WA 98273
Tel: (360) 336-0661
whc@cnw.com
www.highway99.com/whc

Seeds and Seed Cultivars

Ag Innovations
Oxford Hemp Exchange
202 S. Westwood Blvd., Ste. 32
Popular Bluff, MO 63901
Tel: (573) 785-8711
Fax:: 573-785-3059
oxhemp@phmo.net
http://www.pbmo.net/oxhemp

Australian Hemp Resource and
   Manufacture
15 Belmont Crescent
Paddington Qld 4064
Australia
Tel: 61-07-3369-5925
Fax: 61-07-3368-1255
ahrm@hits.net.au

FIBRO-SEED Ltd.
GATE "Rudolph Fleischmann"
   Agricultural
Research Institute
Kompolt (Heves) H-3356 Hungary
Tel/Fax: 36-36-489-000

HempFlax B.V. (see Processors)

Hempseed Organics Ltd.
P.O. Box 11797
London N15 6NQ
Tel/Fax: 44-171-833-3178
hempseed@gn.apc.org

Ohio Hempery (see Distributors)

VIPPO
5400 Adjud (Vrancea) Romania
str. Al. I. Cuza nr. 16

Textiles

Flax Craft
2 Phelps AvenueP.O. Box 797
Tenafly, NJ 07670
Tel: (201) 569-3438
Fax: (201) 569-6780

Heavytex, Inc.
6726 Szeged
Alsokikoto Sor 11
Hungary
Tel: 36-62-435-138
Fax: 36-62-435-143

Hemp Textiles International, Inc.
3200 30th Street
Bellingham, WA 98225-8360
Tel: (360) 650-1684
Fax: (360) 650-0523
hti@cantiva.com
www.cantiva.com

Hemp Traders
2132 Colby Avenue #5
Los Angeles, CA 90025
Tel: (310) 914-9557
Fax: (310) 478-2108
hemptrader@aol.com

Naturetex International
Van Diemenstraat 192
1013 CP Amsterdam
The Netherlands
Tel: 31 (0) 20-420-30-40

Ohio Hempery (see Distributors)

Pickering International
888 Post Street
San Francisco, CA 94109
Tel: (415) 474-2288
Fax: (415) 474-1617
jpickering@compuserve.com

Tall Grass Hemp Company
873 Beatty St., Ste. 401
Vancouver, BC V6B 2M6
Canada
Tel: (800) 616-5900 or
    (604) 488-0109
Fax: (604) 488-0891
tgrass@skybus.com

Videos

The Hempen Road Film Project
2103 Harrison NW, Ste. 2756
Olympia, WA 98502
Tel: (360) 705-3804
uncleweed@olywa.net
www.hempenroad.com

BIBLIOGRAPHY

Allen, James Lane. *The Reign of Law: A Tale of the Kentucky Hemp Fields.* London, England: Macmillan Company, 1900.

Bócsa, Dr. Ivan and Michael Karus. *Cultivation of Hemp.* C.F. Mueler Verlag, Heidelberg, Germany (German edition), 1997. Sebastopol, California: HEMPTECH (English version), 1998.

Borth, Christy. *Modern Chemists and Their Work.* New York: The New Home Library, January 1945.

Clarke, Robert C. "The Cultivation and Use of Hemp in Ancient China." *The Journal of the International Hemp Association* (July 1995): 26–30.

Colborn, Theo, Dianne Dumanoski, and John Peterson Myers. *Our Stolen Future: Are We Threatening Our Fertility, Intelligence, and Survival? A Scientific Detective Story.* New York: Dutton, 1996.

Dempsey, James M. *Fiber Crops.* Gainesville: University Press of Florida, 1975.

Dewey, L.H., and J.L. Merrill. "Hemp Hurds: A Paper-making Material." USDA Bulletin No. 404. Washington, D.C.: U.S. Government Printing Office, October 14, 1916.

Erasmus, Udo. *Fats that Heal, Fats that Kill: The Complete Guide to Fats, Oils, Cholesterol, and Human Health.* Burnaby, B.C.: Alive Books, 1993.

Graedel, T.E., and B.R. Allenby. *Industrial Ecology.* Englewood Cliffs, N.J.: Prentice Hall, 1995

Hart, S. L. "Beyond Greening: Strategies for a Sustainable World." *Harvard Business Review* 75.1 (1997): 66+.

Hawken, Paul. *The Ecology of Commerce: A Declaration of Sustainability.* New York: HarperBusiness, 1993.

Hawken, Paul, Amory Lovins, and Hunter Lovins. *Natural Capitalism: The Coming Efficiency Revolution.* Forthcoming, Hyperion, 1998.

Herer, Jack, Mathias Brockers, and Michael Karus. *Hanf.* Frankfurt, Germany: Zweitausendeins, 1993.

Hopkins, James F. *A History of the Hemp Industry in Kentucky.* Lexington: University of Kentucky Press, 1951.

Houtman, Carl. "Market Analysis for Hemp Fiber as a Feed Stock for Papermaking." Madison, Wis.: USDA Forest Products Laboratory, 1997.

Iowa Agricultural Experiment Station. *Hemp: A War Crop for Iowa.* By C. P. Wilsie, E. S. Dyas, and A. G. Norman. Washington. D.C.: GPO, 1942.

Jones, Kenneth. *Nutritional & Medicinal Guide to Hemp Seed.* Vancouver, British Columbia: Rainforest Botanical Laboratories, 1996.

Kolander, Cheryl. *Hemp for Textile Artists.* Portland, Oregon: MAMA D.O.C., 1995.

Lefsher, Marc. "Hemp's Mantra: It isn't pot, it isn't pot, it isn't pot . . ." *Wall Street Journal,* October 30, 1996.

Lloyd, Irwin, et al. *Ag Fiber Composites: A Comprehensive Report on the Use of Agricultural Crops to Manufacture Building Materials.* Sebastopol, Calif.: HEMPTECH, 1995.

Lotz, Lap, R. Groeneveld, B. Habekotte, and H. Van Oene. "Reduction of Growth and Reproduction of *Cyperus esculentus* by Specific Crops." *Weed Research* 31.3 (1991): 153-160.

McPartland, John. "Cannabis Pests." *The Journal of the International Hemp Association* (December 1996): 49, 52–55.

———. "Cannabis Diseases." *The Journal of the International Hemp Association* (June 1996): 19–23..

M'Gonigle, Michael R., and Ben Parfitt. *Forestopia: A Practical Guide to the New Forest Economy.* Madeira Park, B.C.: Harbour Pub., 1994.

Mintz, John. "Splendor in the Grass?" *Washington Post,* January 5, 1997.

Morris, David J., and Irshad Ahmed. *The Carbohydrate Economy: Making Chemicals and Industrial Materials from Plant Matter.* Washington D.C.: Institute for Local Self-Reliance, 1992.

nova Institute, et al. *Bioresource Hemp: Proceedings of the Symposium.* Sebastopol, Calif.: HEMPTECH, 1995.

Perkins, John M. *The World is as You Dream It: Shamantic Teachings from the Amazon and Andes.* Rochester, N.Y.: Destiny Books, 1994.

Perlin, John. *A Forest Journey: The Role of Wood in the Development of Civilization.* New York: W. W. Norton, 1989.

Reichert, Gordon. *Hemp.* Winnipeg, Manitoba: Agriculture and Agri-Food Canada, 1994.

Riddlestone, Sue, and Desai Pooran. *Bioregional Fibers.* Surrey, England: Bioregional Development Group, 1994.

Robinson, Jeffrey. *The Laundrymen: Inside Money Laundering, the World's Third-Largest Business.* New York: Arcade Pub., 1996.

Roulac, John W. *Industrial Hemp: Practical Products—Paper to Fabric to Cosmetics.* Sebastopol, Calif.: HEMPTECH, Second edition, 1996.

Schafer, E.R., and F.A. Simmonds. "A Comparison of the Physical and Chemical Characteristics of Hemp Stalks and Seed Flax Straw." Publication no. R868. Madison, Wis.: USDA Forest Products Laboratory, 1926

Small, Ernest. *The Species Problem in Cannabis: Science and Semantics.* Toronto: Corpus, 1979.

United States Department of Agriculture, Division of Statistics. *A Report on Flax, Ramie, and Jute: With Considerations Upon Flax and Hemp Culture in Europe, a Report on the Ramie Machine Trials of 1889 in Paris, and Present Status of Fiber Industries in the United States.* By Charles R. Dodge. Washington, D.C.: GPO, 1890.

United States Department of Agriculture. *1944 Hemp Production Experiments: Cultural Practices & Soil Requirements.* Bulletin P-63. Ames, Iowa: Agricultural Experiments Station, 1944.

van der Werf, Hayo. *Crop Physiology of Fiber Hemp.* Amsterdam, the Netherlands: International Hemp Association, 1994.

Wirtshafter, Don. *Schlichten Papers.* Guysville, Ohio: Ohio Hempery, 1994.

Major Cotton Pesticides in the United States

Pesticide chemical name (trade name)	Agricultural use	Immediate toxicity	Long Term toxicity	Environmental toxicity
Aldicarb (Temik)	insects & nematodes	high	cancer, suspect of causing mutations	fish
Chlorpyrifos (Lorsban)	insects	moderate-high	brain and fetal damage, impotence, sterility	amphibians, aquatic insects, bees, birds, crustaceans
Cyanazine (Bladex)	weeds	moderate-high	birth defects, cancer	bees, birds, crustaceans, fish
Dicofol (Kelthane)	mites, & has insecticidal properties	moderate-high	cancer, reproductive damage, tumors	aquatic insects, birds, fish
Ethephon (Prep)	plant growth regulator	moderate	mutations	birds, bees, crustaceans, fish
Fluometuron (Higalcoton)	herbicide	unknown	blood, spleen	bees & fish
Metam Sodium (Vapam)	insects, nematodes, fungus, & weeds	moderate-high	birth defects, fetal damage, mutations	birds & fish
Methyl Parathion (Parathion, Metaphos)	insects	very high	birth defects, fetal damage, immune system & reproductive damages, mutations	bees, birds, crustaceans, fish
MSMA (Mesamate)	herbicide	moderate-high	tumors	bees & fish
Naled (Dibrom) tions,	insects & has miticidal bees, birds,	very high properties	cancer, reproductive damage, suspect of tumors	amphibians, aquatic insects, causing muta- crustaceans, fish
Profenofos (Curacron)	insects & mites	high	eye damage, skin irritant	birds, bees, fish
Prometryn (Primatol Q)	herbicide	moderate-high	bone marrow, kidney, liver, testicular damage	bees, birds, crustaceans, fish, & molusks
Propargite (Omite)	miticide	moderate-high	cancer, fetal & eye damage, mutations tumors	bees, birds, crustaceans, fish
Sodium Chlorate (Fall)	leaf drop & weeds	low	kidney damage & methemoglobinemia	birds & fish
Tribufos (DEF, Folex)	leaf drop	moderate-high	cancer, tumors	birds & fish
Trifluralin (Treflan)	herbicide	low-moderate	cancer, fetal damage, suspect mutagen, teratogen	amphibians, aquatic insects, bees, birds, crustaceans, fish

Source: The Sustainable Cotton Project

INDEX

About the Author

JOHN W. ROULAC is an author, an entrepreneur, and a composting and agricultural-fibers advocate. He has researched sustainable living systems for more than fifteen years. In addition to *Hemp Horizons*, he has written *Industrial Hemp: Practical Products—Paper to Fabric to Cosmetics*, and the best-seller *Backyard Composting: Your Complete Guide to Recycling Yard Clippings* (which has sold over half a million copies). Roulac is widely quoted by the media in publications ranging from *Wired* magazine to the *Wall Street Journal*. In 1987 he founded Harmonious Technologies, an internationally recognized leader in the field of home composting. This company has:

— assisted seven hundred government agencies in implementing home-composting programs

— educated fifteen thousand people who have attended the company's composting seminars

— distributed 200,000 compost bins throughout North America

— designed and manufactured, in partnership with Smith & Hawken, the HOME COMPOSTER bin, made of 100-percent recycled plastic

Roulac has founded three environmental organizations: the Pasadena-based Arroyo Seco Council, which in 1990 produced Southern California's largest Earth Day festival, attracting thirty-five thousand attendees; Forests Forever, which sponsored the 1990 California Forest Protection Initiative; and the North American Industrial Hemp Council (NAIHC), a trade association that is working to recommercialize industrial hemp. Roulac currently serves as Secretary on the NAIHC Board of Directors. In 1994, he founded HEMPTECH, the Industrial Hemp Information Network.